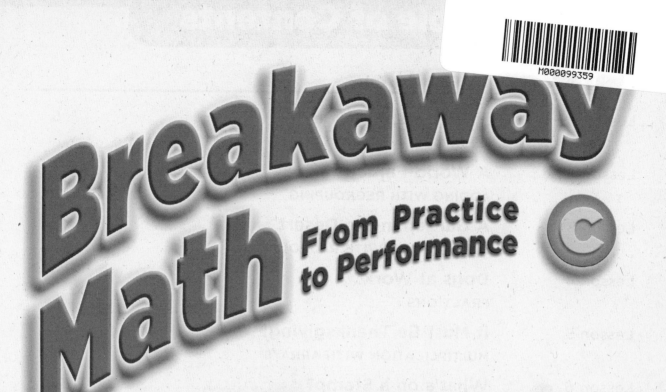

Breakaway Math
From Practice to Performance C

Brian Enright, Ed.D.

Joan Fox

Robert Gyles, Ph.D.

Maxine Leonescu

Fred I. Remer

Options
Publishing Inc.

Toll Free: 800-782-7300
Fax: 866-424-4056
www.optionspublishing.com

Table of Contents

Algebra

Data Analysis and Probability

Math Tools

Acknowledgments

Executive Editor: Linda Bullock

Editor: Joshua Fisher

Production Supervisor: Sandy Batista

Production Specialist: Corrine Scanlon

Design and Production: Design 5 Creatives

Cover Design: Design 5 Creatives

Illustration Credits: Joe Boddy, Molly K. Scanlon

Photo Credits: Cover and title page: Lawrence Migdale, 5: Photos.com, 7: Joseph Van Os/Getty Images, 11 Kevin Schafer/Corbis, 17: Getty Images, 19: Bjorn Backe/Corbis 23: Tom Nebbia/Corbis, 29: Jeff Christensen/Corbis, 31: Bettmann/Corbis, 35: USPS, 37: USPS, 41: Charles O'Rear/Corbis, 47: Photos.com, 53: Lucas Sheridan, 59: Photos.com, 65: David A. Northcott/Corbis, 71: Bettmann/Corbis, 77: Quarter-dollar coin image from the United States Mint, 79: Nickel coin image from the United States Mint, 83: Getty Images, 89: Photos.com, 95: Bettmann/Corbis, 97: Mike Blake/Corbis, 101: Bettmann/Corbis, 103: Corbis, 107: Warren Morgan/Corbis,109: Rockhouse Museum, 113: Getty Images 119: NASA, 125: Getty Images, 127: Trent Nelson/The Salt Lake Tribune, 131: Roger Ball/Corbis, 133: Juan Silva/Getty Images

ISBN-10: 1-59137-372-7

ISBN-13: 978-1-59137-372-8

© 2005 Options Publishing, Inc.

Printed in the U.S.A.

15 14 13 12 11 10

HPS223049

Little Cats, Big Cats

When most people think of cats, they think of cute and cuddly pets. Have you ever thought about their larger relatives—cheetahs, lions, and tigers? While a pet cat may dash through the house very quickly, a cheetah can reach speeds over 65 miles per hour! Your pet cat may seem big at 10 pounds, but a Siberian tiger can weigh more than 800 pounds.

Get Started

A typical housecat weighs about 8 pounds. On the **number line** below, 8 comes between 0 and 10. Notice that 8 is closer to 10 than it is to 0.

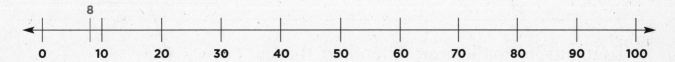

- A cheetah may weigh about 98 pounds. Which pair of tens does 98 come between?

  between _____ and _____

  Show where 98 belongs on the number line.

- A mountain lion may weigh about 75 pounds. Which pair of tens does 75 come between?

  between _____ and _____

  Show where 75 belongs on the number line.

- A lynx may weigh about 43 pounds. Which pair of tens does 43 come between?

  between _____ and _____

  Show where 43 belongs on the number line.

Use a Number Line to Round

You can use a number line to **round** a number to the nearest ten.

To round 23 to the nearest ten, think:

- 23 is between 20 and 30.

- 23 is closer to 20 than it is to 30.

- So 23 rounds down to 20.

You can also use a number line to round a number to the nearest hundred.

Remember
If a number is exactly in the middle, round up: 25 rounds up to 30, and 350 rounds up to 400.

To round 360 to the nearest hundred, think:

- 360 is between 300 and 400.

- 360 is closer to 400 than it is to 300.

- So 360 rounds up to 400.

Use the number lines above to round each number.

1. Round 27 to the nearest ten.

27 rounds to _____.

2. Round 330 to the nearest hundred.

330 rounds to _____.

Use the number lines below to round each number.

3. Round 61 to the nearest ten.

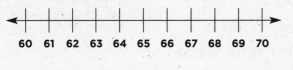

61 rounds to _____.

4. Round 120 to the nearest hundred.

120 rounds to _____.

5. Round 54 to the nearest ten.

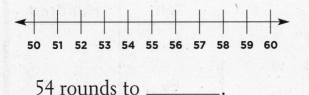

54 rounds to _____.

6. Round 270 to the nearest hundred.

270 rounds to _____.

7. Round 38 to the nearest ten.

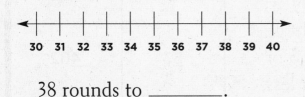

38 rounds to _____.

8. Round 590 to the nearest hundred.

590 rounds to _____.

Solve a Problem

9. Look at the number line in Question 8 above.

• Which pair of tens does 543 come between?

between ___540___ and _____

• What is 543 rounded to the nearest ten? _____

It's a Fact!
A group of lions is called a pride.

Rounding Without Number Lines

Even though the lion is only the second largest of the big cats, a male lion can weigh more than 400 pounds!

Suppose the adult lion shown at the right weighs 462 pounds. His cub weighs 53 pounds. What is the cub's weight rounded to the nearest ten? What is the adult's weight rounded to the nearest hundred?

Follow the steps below to find out.

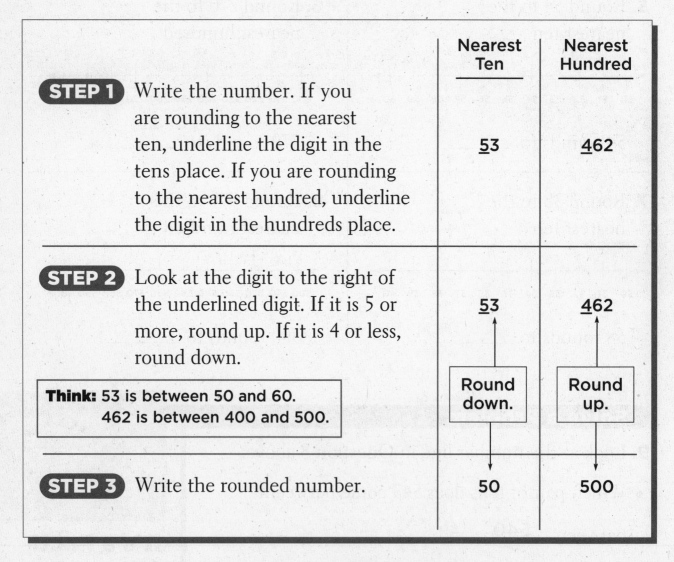

		Nearest Ten	Nearest Hundred
STEP 1	Write the number. If you are rounding to the nearest ten, underline the digit in the tens place. If you are rounding to the nearest hundred, underline the digit in the hundreds place.	5<u>3</u>	<u>4</u>62
STEP 2	Look at the digit to the right of the underlined digit. If it is 5 or more, round up. If it is 4 or less, round down.	5<u>3</u> ↑ Round down. ↓	<u>4</u>62 ↑ Round up. ↓
	Think: 53 is between 50 and 60. 462 is between 400 and 500.		
STEP 3	Write the rounded number.	50	500

The adult lion's weight is between 460 and 470 pounds. What is

the adult's weight rounded to the nearest ten? _____ pounds

Follow the steps you learned to round each number.

1. Round to the
nearest hundred.

231 ⟶ _____

2. Round to the
nearest ten.

93 ⟶ _____

3. Round to the
nearest hundred.

347 ⟶ _____

Round to the nearest ten.

347 ⟶ _____

4. Round to the
nearest hundred.

864 ⟶ _____

Round to the nearest ten.

864 ⟶ _____

5. Round to the
nearest hundred.

750 ⟶ _____

6. Round to the
nearest ten.

15 ⟶ _____

7. Use the number line below to round
each number to the nearest hundred.
Then add the rounded numbers.

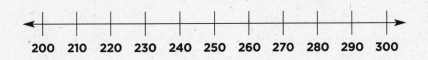

200 210 220 230 240 250 260 270 280 290 300

233 ⟶ _____

267 ⟶ + _____

_____

On Your Own

When you round these
numbers to the nearest ten,
you get 100.

Which numbers are they?

When you round these
numbers to the nearest ten,
you get 0.

Which numbers are they?

1 Test Yourself

1. Round 54 to the nearest ten.

Ⓐ   Ⓑ   Ⓒ   Ⓓ

2. Round 357 to the nearest hundred.

Ⓕ   Ⓖ Ⓗ   Ⓙ

3. When Josh's weight is rounded to the nearest ten, it is 90 pounds. Which of these amounts could be Josh's weight?

Ⓐ 100 pounds

Ⓑ 95 pounds

Ⓒ 85 pounds

Ⓓ 79 pounds

4. What is 464 rounded to the . . .

nearest ten? _____

nearest hundred? _____

5. Marty says he has about 50 CDs in his collection. Marty rounded to the nearest ten. How many CDs could Marty have in his collection? List all the possibilities.

_____

_____

_____

_____

_____

6. Think Back Fill in the blanks.

25 = _____ tens _____ ones

39 = _____ tens _____ ones

70 = _____ tens _____ ones

A World-Famous Zoo

Do you enjoy going to the zoo? A wonderful zoo to visit is the San Diego Zoo in California. But this is no ordinary zoo. The animals at the San Diego Zoo live in their natural habitats rather than in cages. Visitors can see gorillas in a tropical setting, polar bears diving into the water, and monkeys in the rainforest. They can even see koala bears and rare giant pandas.

Get Started

Ramón and Linda made a table to show the number of different animals they saw on a class trip to the zoo. Use the table to answer the question below.

Animals Seen by Ramón and Linda

Animal	Amount
Bears	22
Monkeys	34
Birds	145
Reptiles	47
Bats	112

Show your work here.

- All together, how many bears and reptiles did Ramón and Linda see?

_____

Use Blocks to Regroup

Suppose you want to find the number of birds and reptiles Ramón and Linda saw. You need to add 145 and 47.

Hundreds	Tens	Ones
1	4	5
+	4	7
		12?

Look at the ones column above. When you add 5 and 7, you get 12. But you can't write 12 in the ones column, because only one digit can go in the ones column. So you need to regroup the ones.

Regrouping is like trading. You get different numbers of tens and ones, but the total amount stays the same.

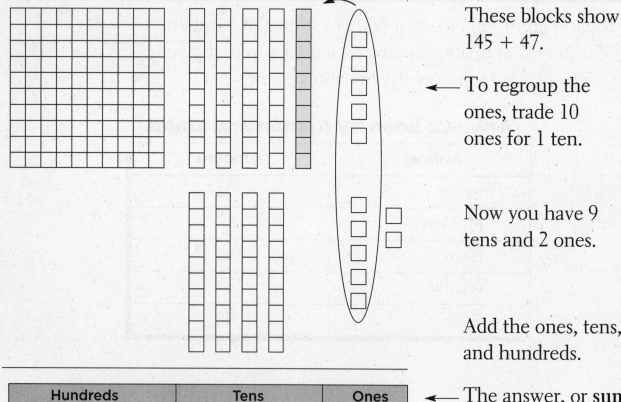

These blocks show 145 + 47.

To regroup the ones, trade 10 ones for 1 ten.

Now you have 9 tens and 2 ones.

Add the ones, tens, and hundreds.

The answer, or **sum,** is 192.

Hundreds	Tens	Ones
1	9	2

1. In the example above, 12 ones was

regrouped as _____ ten and _____ ones.

Practice

Find each sum. Use the drawings to regroup the ones.

2.

Hundreds	Tens	Ones
	6	5
+	2	6

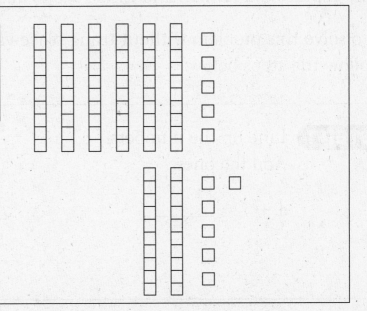

3.

Hundreds	Tens	Ones
1	4	8
+	3	2

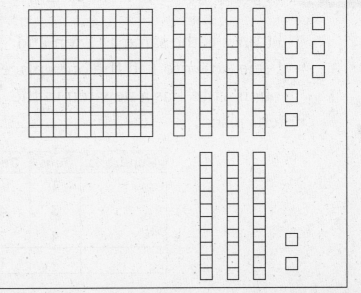

Solve a Problem

4. A total of 143 students and 38 parents went to the zoo. How many people in all went to the zoo?

In all, _____ people went to the zoo.

It's a Fact!
Giant pandas eat mostly bamboo.

Lesson 2: Adding with Regrouping   **13**

Adding with Regrouping

Look back at the table on page 11. How many monkeys and reptiles did Ramón and Linda see in all?

To solve this problem without using place-value blocks, follow the steps below:

STEP 1 Line up the numbers. Add the ones.

$4 + 7 = $ _____

Hundreds	Tens	Ones
	3	4
+	4	7

STEP 2 Regroup the ones if necessary.   11 ones = 1 ten + 1 one

11 ones is the same as 1 ten and 1 one, so write 1 in the ones place. Then write 1 as a new ten in the tens place.

Hundreds	Tens	Ones
	1	
	3	4
+	4	7
		1

Regroup 11 ones as 1 ten and 1 one.

STEP 3 Add the tens. Then add the hundreds if there are any.

Write the sum.

Hundreds	Tens	Ones
	1	
	3	4
+	4	7
		1

Follow the steps you learned to solve these addition problems. Show your work.

1.

Hundreds	Tens	Ones
	2	6
+	3	5

2.

Hundreds	Tens	Ones
1	0	3
+	5	9

3.

Hundreds	Tens	Ones
1	2	5
+	6	5

4.

Hundreds	Tens	Ones
1	5	3
+	3	8

5.

Hundreds	Tens	Ones
2	2	8
+	4	5

6.

Hundreds	Tens	Ones
	8	6
+		8

7. A zoo has 115 birds. It adds 48 more birds. How many birds are at the zoo now? Show your work in the space below.

There are _____ birds at the zoo now.

On Your Own

Find the sum.

$$\begin{array}{r} 159 \\ 114 \\ +\ 221 \\ \hline \end{array}$$

Show your work.

② Test Yourself

1. Which shows how 3 tens and 14 ones can be regrouped?

Ⓐ 2 tens and 4 ones

Ⓑ 4 tens and 0 ones

Ⓒ 4 tens and 4 ones

Ⓓ 3 tens and 4 ones

2. Which number does the picture below represent?

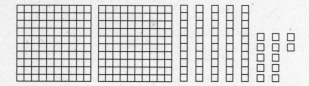

Ⓕ 252

Ⓖ 262

Ⓗ 362

Ⓙ 2,512

3. Find the sum:

$$68 \\ + 26$$

Ⓐ 94

Ⓑ 84

Ⓒ 814

Ⓓ 42

4. Add: 234 + 59 = _____

Why did you have to regroup in this problem?

_____

_____

_____

5. Barry had 327 baseball cards in his collection. Marie had 165 cards in her collection. How many cards did they have in all? Show your work.

They had _____ cards in all.

6. Think Back What is 327 rounded to the nearest . . .

ten? _____

hundred? _____

A Garden in the Desert

Have you ever seen a soaptree yucca or a fairy duster? These plants grow in the deserts of Mexico and parts of the U.S. In summer, a tall stalk grows in the middle of the yucca. Creamy white flowers cover the top half of the stalk. Fairy dusters bloom in the winter and after heavy rains. Other desert plants bloom after rains, too, painting the desert floor purple, red, yellow, and orange.

Get Started

After visiting a garden museum, Mr. Chan's and Ms. Lopez's classes decided to plant gardens at their school. The tables show how many plants each class planted. Use the tables to answer the question below.

Mr. Chan's Class

Plant	Number Planted
Desert Marigold	165
Poppy	48
Cactus	24

Ms. Lopez's Class

Plant	Number Planted
Desert Marigold	52
Poppy	41
Cactus	12

Show your work here.

- How many more desert marigolds did Mr. Chan's class plant than Ms. Lopez's class?

Use Blocks to Regroup

Hundreds	Tens	Ones
1	6	5
−	4	8
		?

Suppose you are in Mr. Chan's class, and you want to know how many more desert marigolds you planted than poppies. You need to subtract: 165 − 48.

Look at the ones column above. You can see that 165 has 5 in the ones place, and 48 has 8 in the ones place. Before you can subtract, the number in the ones place in 165 must be greater than or equal to the number in the ones place in 48.

To get more ones, you need to regroup the tens. Regrouping is like trading one thing for another of equal value.

The blocks below show how to regroup the tens in 165.

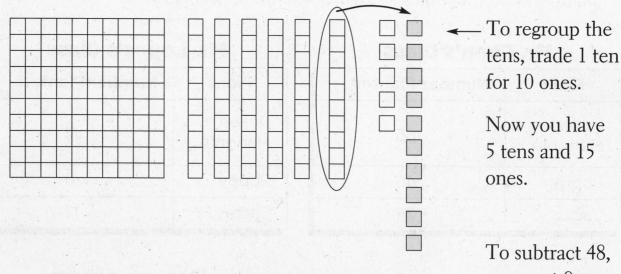

← To regroup the tens, trade 1 ten for 10 ones.

Now you have 5 tens and 15 ones.

To subtract 48, cross out 8 ones. Then cross out 4 tens.

Hundreds	Tens	Ones

← Write the answer, or **difference.**

Practice

Find each difference. Use the drawings
to regroup the tens.

1.

Hundreds	Tens	Ones
1	5	2
−	3	5

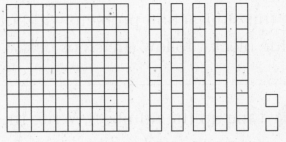

2.

Hundreds	Tens	Ones
1	3	5
−	1	8

Solve a Problem

3. Vicky has 141 baseball cards in her collection.
Jared has 119 baseball cards. How many more
baseball cards does Vicky have than Jared?
Use the space below to show your work.

It's a Fact!

The saguaro cactus
is the tallest cactus
in the world. It can grow
up to 67 feet high!

Vicky has _____ more baseball cards
than Jared.

Subtracting with Regrouping

Suppose there are 143 desert marigolds and cactus plants in all. Of these plants, 29 are cactus plants. How many of the plants are desert marigolds?

To solve this subtraction problem without using place-value blocks, follow the steps below.

STEP 1 Line up the numbers. Write the greater number on top.

Look at the ones column: 3 is less than 9, so you need to regroup.

Hundreds	Tens	Ones
1	4	3
	2	9

−

STEP 2 Regroup the tens. 4 tens = 3 tens + 10 ones

4 tens is the same as 3 tens + 10 ones, so cross out the 4 in the tens column and write 3. Then add the 10 ones to the 3 ones, and write 13 in the ones column.

Hundreds	Tens	Ones
1	³4̶	¹³3̶
	2	9

−

Trade 1 ten for 10 ones.

STEP 3 Subtract. Write the difference.

13 ones − 9 ones = 4 ones

3 tens − 2 tens = _____ ten

1 hundred − 0 hundreds = _____ hundred

Hundreds	Tens	Ones
1	³4̶	¹³3̶
	2	9
		4

−

Follow the steps you learned to solve these
subtraction problems. Show your work.

1.

Hundreds	Tens	Ones
1	8	7
	5	8

2.

Hundreds	Tens	Ones
2	6	2
1	3	7

3.

Hundreds	Tens	Ones
1	3	1
1	1	3

4.

Hundreds	Tens	Ones
	5	8
	4	9

5.

Hundreds	Tens	Ones
	7	5
	2	6

6.

Hundreds	Tens	Ones
2	6	4
1	2	5

7. The school auditorium seats 275 people.
If there are 48 third graders seated in the
auditorium, how many seats are left?
Show your work in the space below.

There are _____ seats left.

On Your Own

Tim solved this
subtraction problem:

$$375$$
$$-\ 248$$
$$\overline{133}$$

Use addition to show why
Tim's answer is incorrect.
Then subtract to find the
correct answer.

1. Which shows how 5 tens and 3 ones can be regrouped?

   Ⓐ 6 tens and 13 ones

   Ⓑ 5 tens and 13 ones

   Ⓒ 4 tens and 13 ones

   Ⓓ 3 tens and 13 ones

2. Subtract: $34 - 15 = \square$

   Ⓕ 49

   Ⓖ 29

   Ⓗ 21

   Ⓙ 19

3. Find the difference:

$$\begin{array}{r} 582 \\ -\ 48 \\ \hline \end{array}$$

   Ⓐ 534

   Ⓑ 536

   Ⓒ 546

   Ⓓ 608

4. Erica's goal is to do 75 pushups. So far she has done 59. How many more does she have to do? Show your work.

   Erica has _____ more pushups to do.

5. Subtract: $267 - 28 =$ _____ Show your work.

6. **Think Back** Find the sum. Show your work below.

   $45 + 138 =$ _____

Dogs at Work

Many dogs are trained to do jobs that help people. Some dogs are trained to guide people who cannot see. Huskies can be trained to pull sleds across snow-covered areas where cars and trucks cannot go. Saint Bernards have been trained to find and rescue people who are lost in the mountains. It is no wonder that dogs have been called our "best friends."

Get Started

At the Cute Pooch Puppy Place, several kinds of terriers are raised as pets. The diagrams below show the kennel space for each type of dog. Use the diagrams to answer the questions.

- How many equal parts does this kennel have?

_____

- What fractional part of this kennel do the Jack Russell terriers live in?

_____

Jack Russell Terriers		

- How many equal parts does this kennel have?

_____

- What fractional part of this kennel does **not** have fox terriers living in it?

_____

Fox Terriers	
Fox Terriers	Fox Terriers

Add and Subtract Fractions with Like Denominators

In a fraction, the bottom number is called the **denominator.**
You can use fraction models to add and subtract fractions
with the same denominator.

To find $\frac{2}{6} + \frac{3}{6}$,

Use a model with six equal
parts. Shade $\frac{2}{6}$ of the model.

Using a different color, shade
$\frac{3}{6}$ more of the model.

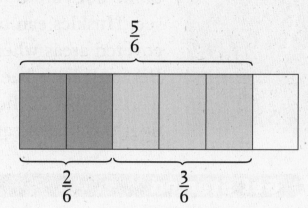

Now count how many parts are
shaded in all. Then solve the problem: $\frac{2}{6} + \frac{3}{6} = \frac{5}{6}$.

To find $\frac{5}{6} - \frac{3}{6}$,

Use a model with six equal
parts. Shade $\frac{5}{6}$ of the model.

Cross out $\frac{3}{6}$ of the model to show
that 3 equal parts were taken away.

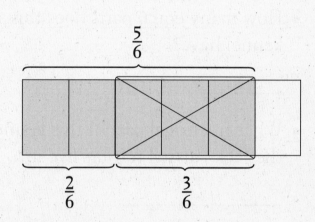

Now count how many parts are left.
Then solve the problem: $\frac{5}{6} - \frac{3}{6} = \frac{2}{6}$.

1. The model at the right has 8 equal parts.
Use the model to add and subtract below.

$\frac{1}{8} + \frac{3}{8} =$ _____

$\frac{4}{8} - \frac{3}{8} =$ _____

Practice

Shade each fraction model below to show the number sentence on the right. Write each sum or difference.

2.

$$\frac{3}{8} + \frac{4}{8} = \boxed{}$$

3.

$$\frac{7}{12} + \frac{3}{12} = \boxed{}$$

4.

$$\frac{3}{5} - \frac{1}{5} = \boxed{}$$

5.

$$\frac{5}{8} - \frac{2}{8} = \boxed{}$$

Solve a Problem

6. Michelle is making brownies. She uses $\frac{1}{4}$ of a stick of butter for the batter and $\frac{2}{4}$ of the stick for the icing. How much of the stick of butter was used in all? Draw a model and write a number sentence.

It's a Fact!

Long ago, Barry, a famous Saint Bernard, rescued more than 40 people lost in the snowy mountains of the Alps in Switzerland.

Lesson 4: Fractions   **25**

Add and Subtract Fractions

In a **fraction,** the top number is called the **numerator.** The bottom number is called the **denominator.** Follow the steps below to add and subtract fractions without models.

$\dfrac{1}{3}$ ←— numerator
←— denominator

Suppose you want to find this sum: $\dfrac{1}{3} + \dfrac{1}{3} = \boxed{}$

STEP 1   Write the denominator in the answer.

The denominator in the answer is the same denominator as in the fractions.

$$\dfrac{1}{3} + \dfrac{1}{3} = \dfrac{\square}{3}$$

STEP 2   Add the numerators.

The numerator in the answer is the sum of the numerators in the fractions.

$$\dfrac{1}{3} + \dfrac{1}{3} = \dfrac{2}{3}$$

Suppose you want to find this difference: $\dfrac{3}{4} - \dfrac{2}{4} = \boxed{}$

STEP 1   Write the denominator in the answer.

The denominator in the answer is the same denominator as in the fractions.

$$\dfrac{3}{4} - \dfrac{2}{4} = \dfrac{\square}{4}$$

STEP 2   Subtract the numerators.

$$\dfrac{3}{4} - \dfrac{2}{4} = \dfrac{1}{4}$$

Show What You Know

Follow the steps you learned to add or subtract the fractions below.

1. $\frac{1}{6} + \frac{4}{6} =$ ☐

2. $\frac{1}{4} + \frac{1}{4} =$ ☐

3. $\frac{3}{8} + \frac{2}{8} =$ ☐

4. $\frac{3}{5} - \frac{2}{5} =$ ☐

5. $\frac{7}{8} - \frac{5}{8} =$ ☐

6. $\frac{3}{4} - \frac{2}{4} =$ ☐

Complete the number sentences below.

7. $\frac{1}{12} +$ ☐ $= \frac{7}{12}$

8. ☐ $+ \frac{4}{10} = \frac{7}{10}$

9. ☐ $- \frac{2}{9} = \frac{2}{9}$

10. When you add fractions with the same denominators, why do you only add the numerators? Explain.

_____

_____

_____

_____

On Your Own

Solve:

$\frac{5}{8} + \frac{1}{8} - \frac{3}{8} =$ ☐

List the steps you followed to solve the problem.

1. Which fraction names the shaded part?

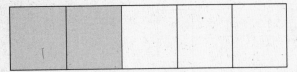

Ⓐ $\frac{1}{5}$ Ⓒ $\frac{3}{5}$

Ⓑ $\frac{2}{5}$ Ⓓ $\frac{2}{3}$

2. Find the sum:

$$\frac{3}{7} + \frac{2}{7} = \boxed{}$$

Ⓕ $\frac{5}{7}$ Ⓗ $\frac{5}{14}$

Ⓖ $\frac{6}{7}$ Ⓙ $\frac{6}{14}$

3. Find the difference:

$$\frac{9}{10} - \frac{3}{10} = \boxed{}$$

Ⓐ $\frac{12}{10}$ Ⓒ $\frac{3}{10}$

Ⓑ $\frac{6}{10}$ Ⓓ $\frac{6}{0}$

4. Shade the model below to show the number sentence.

$$\frac{5}{9} + \frac{3}{9} = \boxed{}$$

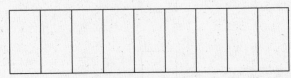

What is the sum? _____

5. Shade the model below to show the number sentence.

$$\frac{11}{12} - \frac{4}{12} = \boxed{}$$

What is the difference? _____

6. Think Back Subtract.

$$\begin{array}{r} 145 \\ -\ 118 \\ \hline \end{array}$$

What is the difference?

_____

It Must Be Thanksgiving!

Have you ever watched the Macy's Thanksgiving Day Parade in New York City? It is one of the most famous parades in the U.S. It includes floats, marching bands, clowns, and giant balloons. Each year, about 2,500,000 people line the city streets to watch the parade. Another 44 million people watch it from home on their televisions.

Get Started

Some students from Greenwood School marched in the Thanksgiving Day Parade. The pictures below show how the third and fourth graders lined up for the parade.

Add the number of students in each row to find how many third graders and fourth graders marched in the parade.

3rd grade

- How many students are in each row? _____
- How many third-grade students marched in all? _____

_____
+ _____
_____

4th grade

- How many students are in each row? _____
- How many fourth-grade students marched in all? _____

_____
_____
_____
+ _____
_____

Using Arrays to Multiply

An **array** is a drawing that shows items in rows. Each row in an array has the same number of items.

The picture below shows a 4 by 5 array. That means that there are 4 rows with 5 items in each row.

There are 4 rows of balloons.

There are 5 balloons in each row.

There are many ways to find the total number of balloons. You can . . .

- Count the balloons, one by one.

  $1, 2, 3, 4, 5, 6, \ldots 20$

- Find the number of balloons in one row.
  Then use repeated addition.

  $5 + 5 + 5 + 5 = 20$

- Multiply. Four rows of 5 is 4 times 5.
  You can write 4 times 5 as 4×5.

  $4 \times 5 = 20$

---

1. Look at the array at the right. Use repeated addition to find the total number of drums. Then **multiply** to find the total number.

_____ + _____ + _____ = _____

___3__ × _____ = _____

Find the total number of items in each array. First,
use repeated addition. Then multiply.

2.

_____ + _____ + _____ = _____

_____ × _____ = _____

3.

_____ + _____ = _____

_____ × _____ = _____

4.

_____ + _____ + _____ = _____

_____ × _____ = _____

Solve a Problem

5. Kwan's band is marching in the parade.
The players are lined up so that there are
4 players in each row. If there are 24
players in Kwan's band, how many rows of
players are there?

_____ rows of 4 players

It's a Fact!
Over 4,000
volunteers help
with the Macy's
Thanksgiving Day
Parade.

Building Arrays

Suppose you want to draw an array for the number sentence $3 \times 5 = ?$

Follow the steps below to build an array.

STEP 1 Look at the number sentence. The first number, or **factor,** tells you how many rows there are.

$3 \times 5 = ?$

There are 3 rows.

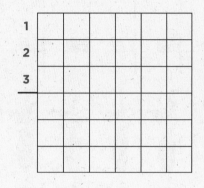

STEP 2 The second factor tells you how many items are in each row.

$3 \times 5 = ?$

There are 5 items in each row.

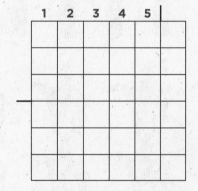

STEP 3 Shade in the first row.

There are 5 items in each row, so there are 5 columns.

STEP 4 Shade in the rest of the rows.

Your array should have 3 rows and 5 columns shaded.

The answer is called the **product.**

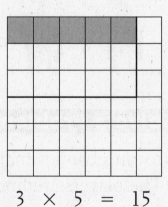

$$3 \times 5 = 15$$

factor   factor   product

Follow the steps you learned to build an array for each of the number sentences below. Then answer the questions that follow.

1. 5 × 5 = _____

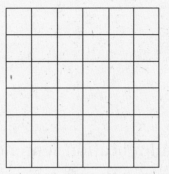

How many rows did you shade? _____

How many columns did you shade? _____

2. 3 × 4 = _____

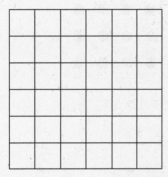

How many rows did you shade? _____

How many columns did you shade? _____

3. 6 × 5 = _____

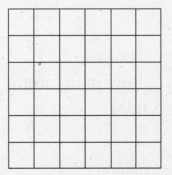

How many rows did you shade? _____

How many columns did you shade? _____

4. What multiplication sentence can you write for this addition sentence: 2 + 2 + 2 + 2?

_____ × _____ = _____

On Your Own

Write five multiplication problems. Then build an array to solve each problem.

1. Which is another way to write
3 + 3 + 3 + 3 + 3?

Ⓐ 5 − 3

Ⓑ 5 ÷ 3

Ⓒ 5 + 3

Ⓓ 5 × 3

2. Which multiplication sentence
does this array show?

Ⓕ 3 × 5 = 15

Ⓖ 6 × 4 = 24

Ⓗ 3 × 6 = 18

Ⓙ 7 × 3 = 21

3. In an array showing
2 × 8 = 16, how many items
are in each row?

Ⓐ 8

Ⓑ 10

Ⓒ 16

Ⓓ 32

4. Draw an array to show that
Chrissie has 3 groups of
seashells with 7 seashells in
each group.

5. Write a multiplication sentence
to describe the array in
Question 4.

_____ × _____ = _____

6. Think Back Find the sum
and the difference. Show your
work below.

116 + 148 = _____

325 − 118 = _____

What's on a Stamp?

Did you know that in the U.S. anyone can suggest ideas for what picture goes on a stamp? People can send their ideas to a special committee. This committee then decides which ideas to share with the head of the post office. Today, you can buy stamps that show special places, like national parks. You can also buy stamps that show famous people and events in history. What would *you* like to see on a stamp?

Get Started

Draw lines in the boxes below to answer each question. The first one has been done for you.

- John bought a sheet of stamps. There are 4 rows and 5 stamps in each row.

  How many stamps are there? __20__

- Ms. Williams buys a sheet of stamps. There are 3 rows and 5 stamps in each row.

  How many stamps are there? _____

  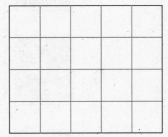

- Mr. Cruz buys a sheet of stamps. There are 2 rows and 4 stamps in each row.

  How many stamps are there? _____

Working with Division

Division is the opposite of multiplication. When you **divide**, you break a group into smaller groups. Each smaller group has an equal number of items.

Look at the picture below. All together, there are 20 stamps. Suppose you divide the stamps into groups of 4.

One group of 4 has been circled for you. Circle the other groups of 4.

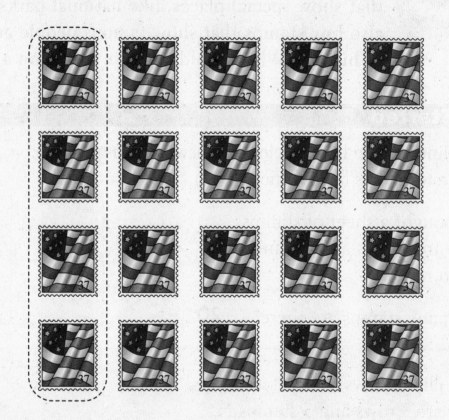

You can write a division sentence to represent this picture.

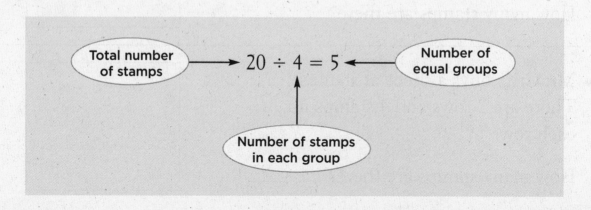

Complete each division sentence below.

1. There are 16 stars.
   Circle groups of 4.

   There are _____ equal groups.

   16 ÷ _____ = _____

2. There are 18 triangles.
   Circle groups of 3.

   There are _____ equal groups.

   18 ÷ _____ = _____

3. Draw a picture at the right
   to show this division sentence:

   14 ÷ 2 = _____

4. Sandra has 24 stamps in her collection.
   She wants to organize the stamps into
   groups of 4. Draw a picture to show how
   many groups Sandra will make.

It's a Fact!

Dr. Martin Luther King, Jr.
was a famous civil rights
leader. Now there is a
stamp in his honor.

Using Arrays to Divide

You know how to use arrays to multiply. You can also use arrays to divide.

Follow the steps below to write a division sentence by using an array.

STEP 1 Count the total number of items in the array. This is the first number in the division sentence. It is called the **dividend.**

$$\underline{\quad 20 \quad} \div \underline{\qquad} = \underline{\qquad}$$

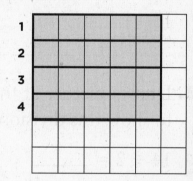

STEP 2 Count the number of rows. This is the second number in the division sentence. It is called the **divisor.**

$$\underline{\quad 20 \quad} \div \underline{\quad 4 \quad} = \underline{\qquad}$$

STEP 3 Count the number of columns. This is the answer, or **quotient.**

$$\underline{\quad 20 \quad} \div \underline{\quad 4 \quad} = \underline{\quad 5 \quad}$$

STEP 4 Turn the array on its side so you see 5 rows and 4 columns. Follow the steps above to write a new division sentence for the array.

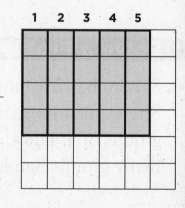

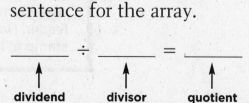

dividend   divisor   quotient

Follow the steps you learned to write two division sentences for each array.

1.

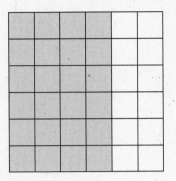

_____ ÷ _____ = _____

_____ ÷ _____ = _____

2.

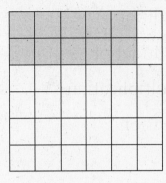

_____ ÷ _____ = _____

_____ ÷ _____ = _____

3.

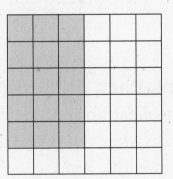

_____ ÷ _____ = _____

_____ ÷ _____ = _____

4.

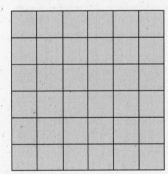

_____ ÷ _____ = _____

_____ ÷ _____ = _____

5. Kira bought a sheet of stamps. The sheet has 6 rows. There are 5 stamps in each row. How many stamps did Kira buy in all?

_____

On Your Own

Look at the arrays above. Write multiplication sentences for each array.

What do you notice?

6 Test Yourself

1. Which operation is the opposite of division?

Ⓐ addition

Ⓑ subtraction

Ⓒ multiplication

Ⓓ division

2. Which division sentence does this array show?

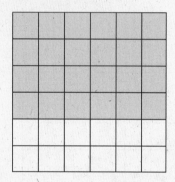

Ⓕ 30 ÷ 6 = 5   Ⓗ 20 ÷ 5 = 4

Ⓖ 18 ÷ 3 = 6   Ⓙ 24 ÷ 4 = 6

3. In art class, 28 crayons were divided into boxes holding 4 crayons each. Which division sentence shows the number of boxes that were used?

Ⓐ 21 ÷ 3 = 7

Ⓑ 28 ÷ 4 = 7

Ⓒ 24 ÷ 3 = 8

Ⓓ 24 ÷ 4 = 6

4. Draw an array that shows 30 ÷ 5.

5. Use the array you drew in Question 4.

What is 30 ÷ 5? _____

6. Think Back Write two multiplication sentences and two division sentences to describe this array.

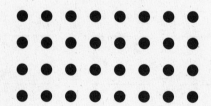

_____ × _____ = _____

_____ × _____ = _____

_____ ÷ _____ = _____

_____ ÷ _____ = _____

Skyscrapers

The Sears Tower in Chicago was finished in 1973. It is 110 stories high, making it the tallest building in North America. Another famous skyscraper is the Transamerica Pyramid in San Francisco. At the top of this building are four cameras—one pointing north, one pointing south, one pointing east, and one pointing west. The cameras show views of San Francisco from 853 feet up.

Get Started

Look at this drawing of a building. List the shapes you can see.

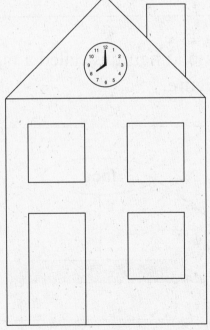

_____

_____

_____

_____

Working with Plane and Solid Figures

The shapes you listed on page 41 are often called **plane figures**. Plane figures are flat shapes that can fit together to make **solid figures**.

Plane figures have only two **dimensions**—length and width. Solid figures have three dimensions—length, width, and height.

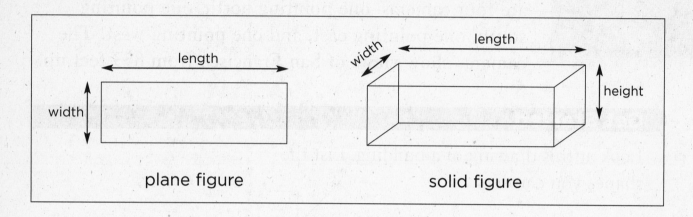

plane figure   solid figure

Each flat surface of a solid figure is called a **face**. Each face is a plane figure.

face a flat surface of a solid figure

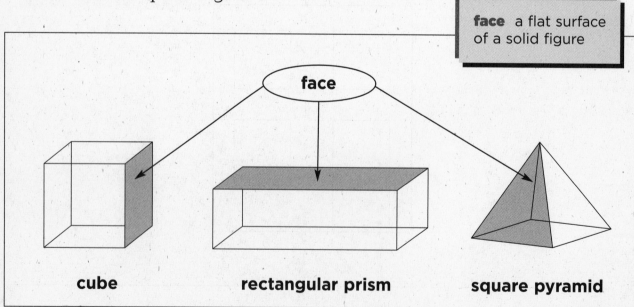

cube   rectangular prism   square pyramid

1. The faces of a cube are all squares. How many square

faces does a cube have? _____

Practice

For <u>Q</u>uestions 2–5, the plane figures that make up
each solid figure are shown. Study each solid figure.
Then answer the questions.

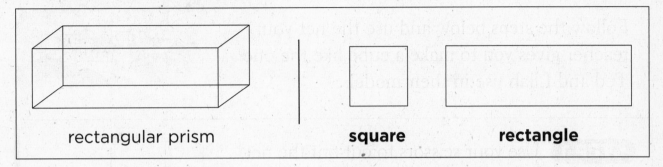

rectangular prism   **square**   **rectangle**

2. How many faces of the solid
figure are rectangles? _____

3. How many faces of the solid
figure are squares? _____

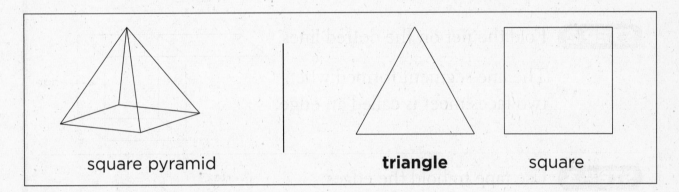

square pyramid   **triangle**   square

4. How many faces of the solid
figure are triangles? _____

5. How many faces of the solid
figure are squares? _____

Solve a Problem

6. How many faces
does this solid
figure have?

_____

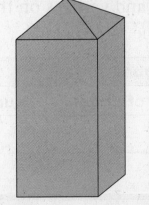

It's a Fact!

From the top of the Empire State
Building in New York City, you can
see New York, New Jersey, Connecticut,
Massachusetts, and Pennsylvania.

Nets of Solid Figures

Ted and Lilah are building a model of their neighborhood. They use a cube as the shape for part of a house in their model.

Follow the steps below and use the net your teacher gives you to make a cube like the one Ted and Lilah use in their model.

STEP 1 Use your scissors to cut out the net.

The **net** shows you how many faces the cube has.

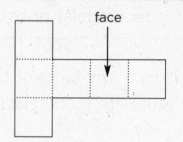

STEP 2 Fold the net on the dotted lines.

The line segment formed when two faces meet is called an **edge**.

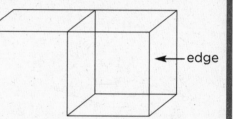

STEP 3 Use tape to hold the edges together to form the solid figure.

The point where three or more edges meet is called a **vertex**. The plural of *vertex* is *vertices*.

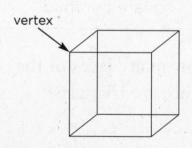

Count the number of faces, edges, and vertices on the cube you made. Write your answers in the table below.

Cube		
Number of Faces	**Number of Edges**	**Number of Vertices**

Show What You Know

Use the steps you learned and nets your teacher
gives you to make each of the solid figures below.
Then fill in the table for each figure.

1.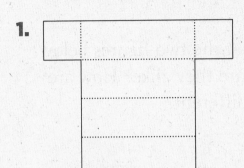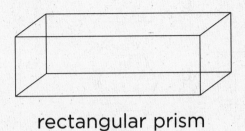

rectangular prism

Rectangular Prism		
Number of Faces	**Number of Edges**	**Number of Vertices**

2.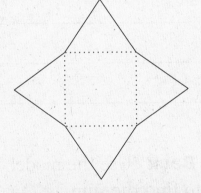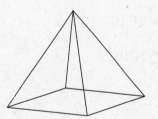

square pyramid

Square Pyramid		
Number of Faces	**Number of Edges**	**Number of Vertices**

3. The point where three or more edges meet is called a _____.

On Your Own

Find a solid figure at home
that looks like a cube or
rectangular prism. Point out
the faces, edges, and vertices.

© 2005 Options Publishing, Inc.

Lesson 7: Solid Figures **45**

7 Test Yourself

1. Which is an example of a solid figure?

Ⓐ 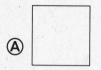 Ⓒ

Ⓑ 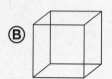 Ⓓ

2. Which figure below is **not** a solid figure?

   Ⓕ cube

   Ⓖ square

   Ⓗ rectangular prism

   Ⓙ square pyramid

3. How many edges does this cube have?

   Ⓐ 4

   Ⓑ 6

   Ⓒ 8

   Ⓓ 12

4. A flat surface of a solid figure is called a _____.

5. Look at the two figures below. How are they alike? How are they different?

_____

_____

_____

_____

6. **Think Back** Use the model below to find the sum.

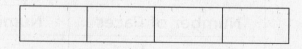

$\frac{1}{4} + \frac{1}{4} =$ ☐

Nature's Halves

What do you see when a butterfly opens its wings? Perhaps you notice that both wings look alike. Some butterfly wings have bright colors. Others have unusual patterns, like dots that look like the eyes of an owl or a snake. So a butterfly's wings do more than help it fly. They can also help protect it from enemies.

Get Started

Look at the pictures below. Draw lines to match each pair of butterfly's wings.

Working with Symmetry

You can fold some shapes so that the two parts of the shape match exactly.

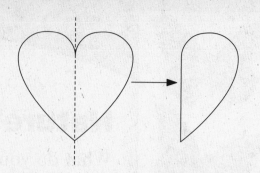

The line you use to fold a shape in half is called a line of symmetry. Look at the lines of symmetry in the drawings below. Notice that some shapes can have more than one line of symmetry.

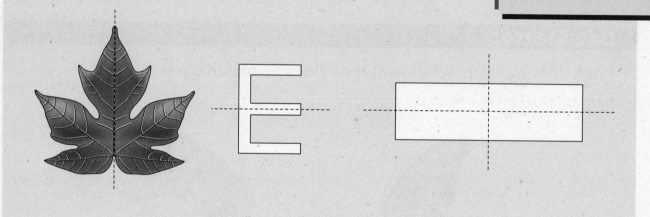

A square has many lines of symmetry. One line of symmetry is shown below.

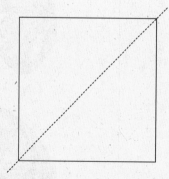

1. Draw the other lines of symmetry for the square.

2. How many lines of symmetry does a square have? _____

Practice

Study each of the shapes below. Try to draw at least one line of symmetry for each shape. Then answer the questions.

3. Were you able to draw a line of symmetry for all of the shapes? Explain your answer.

_____

_____

4. Which shape has more than one line of symmetry?

_____

Solve a Problem

5. Find a shape in your classroom that has at least one line of symmetry. Draw the shape and show its line(s) of symmetry.

It's a Fact!
The fastest butterflies can fly at a speed of about 30 miles per hour!

Constructing Figures with Symmetry

Many things in nature have lines of symmetry. You can make a figure that has a line of symmetry by folding and cutting a piece of paper.

Follow the steps below to make a figure that has a line of symmetry.

STEP 1 Fold a piece of paper in half. Draw a shape on one part of the paper.

Make sure to draw the shape near the fold as shown.

STEP 2 Keeping the paper folded, cut around the shape.

STEP 3 Unfold the paper cutout. The fold is the line of symmetry.

Draw the line of symmetry for the figure above.

Draw each shape below on a sheet of paper folded in half. Then follow the steps you learned to construct a figure with a line of symmetry. Trace each cutout figure in the space at the right and draw a line of symmetry for the figure.

1.

2.

3.

4.

5. Draw a line of symmetry for the figure below.

On Your Own

Find a shape at home that has two or more lines of symmetry. Draw the shape and show its lines of symmetry.

8 Test Yourself

1. How many lines of symmetry does this heart have?

- Ⓐ 0
- Ⓑ 1
- Ⓒ 2
- Ⓓ 3

2. Which letter does **not** have a line of symmetry?

- Ⓕ B
- Ⓖ W
- Ⓗ R
- Ⓙ T

3. Imagine drawing a picture of your foot. How many lines of symmetry would it have?

- Ⓐ 0
- Ⓑ 1
- Ⓒ 5
- Ⓓ 10

4. How many lines of symmetry does this square have? Explain your answer.

_____

_____

_____

_____

5. Look at the equilateral triangle below.

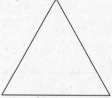

Draw all of the triangle's lines of symmetry.

6. Think Back What is 524 rounded to the nearest hundred?

_____

Legend of the Tangram

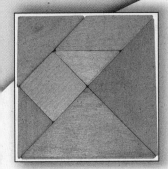

There are many stories about how the first tangram puzzle was made. One story says that in ancient China, a man named Tan dropped a square tile on the floor. The tile broke into 7 geometric pieces. He tried to put the tile back together. That's when he found he could use the pieces to make different shapes and figures.

Get Started

The picture below shows one way you could arrange the seven tangram pieces.

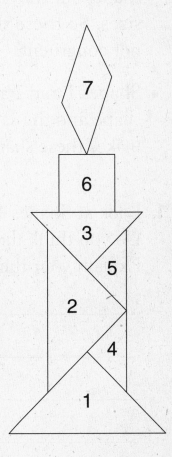

- Look at Shapes 1 and 4. Tell how they are alike and how they are different.

Alike: _____

_____

Different: _____

_____

- Look at Shapes 6 and 7. Tell how they are alike and how they are different.

Alike: _____

_____

Different: _____

_____

Working with Congruent Figures

In geometry, figures that are the same size and the same shape are called **congruent figures**. When two figures are congruent, they will match exactly when one is placed on top of the other.

Look at the tangram square below.

- Shapes 6 and 7 are different shapes. So these shapes are <u>not</u> congruent.

- Shapes 1 and 4 are the same shape, but they are different sizes. So these shapes are <u>not</u> congruent.

- Shapes 1 and 2 are the same shape and the same size. That makes these shapes congruent.

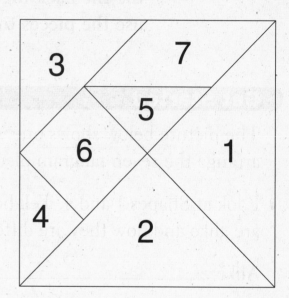

1. Look at Shapes 4 and 5 in the tangram square. Do you think they are congruent? Explain your thinking.

_____

_____

_____

_____

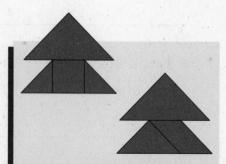

It's a Fact!

One tangram challenge is to use all seven pieces to make two figures that look exactly the same.

Practice

Use your scissors to cut the tangram your teacher gives you into seven pieces. Then use the pieces to answer the questions below.

2. Which two shapes can you use to make a figure congruent to Shape 7?

_____ and _____

3. Which two shapes can you use to make a figure congruent to Shape 6?

_____ and _____

4. Which two shapes can you use to make a figure congruent to Shape 3?

_____ and _____

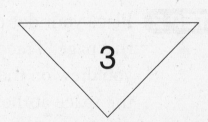

Solve a Problem

5. Together, Shapes 4, 5, and 6 can make this figure:

Choose some of the shapes that are left to draw a figure congruent to the figure above. Show your answer in the space at the right.

Making Congruent Figures

You can use dot paper or a geoboard to make congruent figures. You can also use paper and pencil to trace shapes.

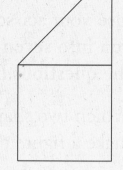

Use dot paper and follow the steps below to make a figure congruent to the figure shown at the right.

STEP 1 Trace the figure onto the dot paper.

Make sure to line up each vertex of the figure with a dot.

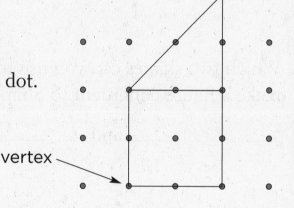

vertex

STEP 2 Place your dot paper under this page. Trace the figure you drew on the dot paper in the space at the right.

The figure you trace is congruent to the figure above.

Use the steps above to draw a figure congruent to the shape shown. Draw the figure in the space to the right of the shape.

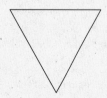

Use the steps you learned to draw a figure
congruent to each shape below.

1.

2.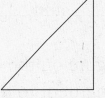

3.

4.

On Your Own

Use all 7 of your tangram
pieces to make an interesting
shape like the one at the left.
Draw an outline of your
shape. Then give your outline
to a friend. Challenge your
friend to figure out how you
built your shape.

9 Test Yourself

1. What do we call two figures that are the same shape and the same size?

   Ⓐ triangles

   Ⓑ squares

   Ⓒ congruent

   Ⓓ symmetry

2. Which figure is congruent to the shape below?

   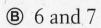

   Ⓕ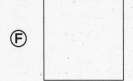

   Ⓖ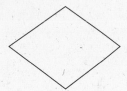

   Ⓗ

   Ⓙ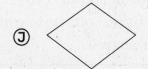

3. Which two tangram pieces listed below are congruent?

   Ⓐ 3 and 4

   Ⓑ 6 and 7

   Ⓒ 1 and 3

   Ⓓ 1 and 2

   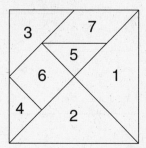

4. Which two other tangram pieces are congruent?

   _____ and _____

5. Draw a figure on the dot paper that is congruent to the figure shown below.

   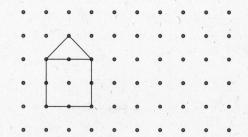

6. **Think Back** Look at the tangram shapes in Question 3 above. Name as many plane figures as you can.

   _____

   _____

Maps and More Maps

There are many kinds of maps. Some maps show stars in the night sky or mountains on the ocean floor. Some show forests, deserts, and rivers. Others show countries and cities. There are also maps that show roads and highways. They help us get from one place to another. You might say we'd be lost without maps!

Get Started

Millie drew this map of her neighborhood. You can see how Millie gets to school each day. First, she walks 1 block east. Then she walks 5 blocks north.

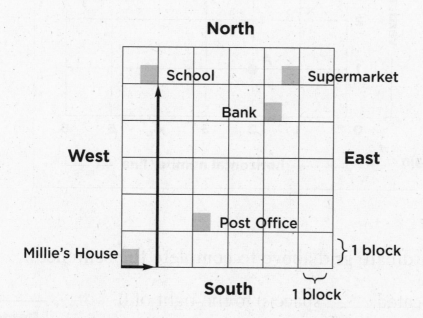

- From her house, Millie walked 2 blocks east and 1 block north. Use a green crayon to trace Millie's path. Where did Millie walk to?

Working with a Coordinate Grid

A **coordinate grid** is like a map. It is used to locate points. A coordinate grid has a **horizontal** number line and a **vertical** number line. The two number lines meet at a point called the **origin**.

The coordinate grid below shows point A and point B. These points are located where two lines cross.

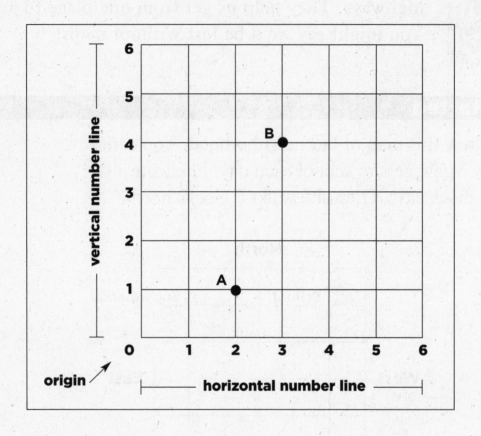

Use the coordinate grid above to complete the sentences.

1. Point A is located _____ space(s) to the right of 0

and _____ space(s) up.

2. Point B is located _____ space(s) to the right of 0

and _____ space(s) up.

> **Remember**
> To locate a point on a coordinate grid, always start at the origin.

Use the coordinate grid below to complete the sentences.

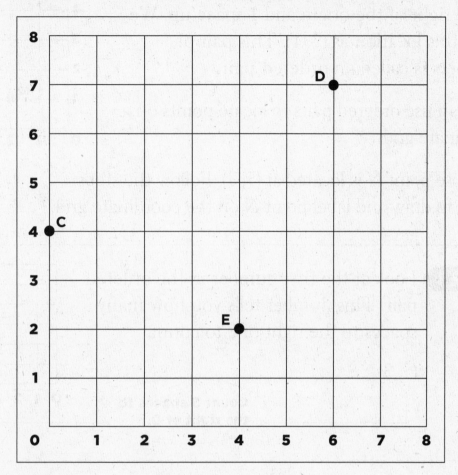

3. Point *C* is located _____ space(s) to the right

of 0 and _____ space(s) up.

4. Point *D* is located _____ space(s) to the right

of 0 and _____ space(s) up.

5. Point *E* is located _____ space(s) to the right

of 0 and _____ space(s) up.

Solve a Problem

6. Draw point *F* on the grid above. Point *F* is
located 5 spaces to the right of 0 and
4 spaces up.

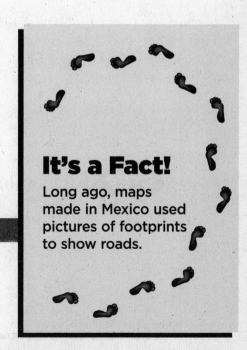

It's a Fact!

Long ago, maps
made in Mexico used
pictures of footprints
to show roads.

Using Ordered Pairs

Look at point M. Point *M* is located 2 spaces to the right of the origin and 1 space up. We write this location as (2, 1). This pair of numbers is called an **ordered pair**.

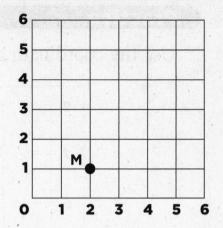

You can use ordered pairs to locate points on a coordinate grid.

Suppose point N is located at (5, 3). Follow the steps below to draw and label point N on the coordinate grid.

STEP 1 Look at the first number in the ordered pair. This number tells you how many spaces to the right of 0 to count.

(5, 3)

Count 5 spaces to the right of 0.

STEP 2 Look at the second number in the ordered pair. This number tells you how many spaces up from 0 to count.

(5, 3)

Count 3 spaces up from 0.

STEP 3 Draw and label the point on the grid.

Point N is located at (5, 3).

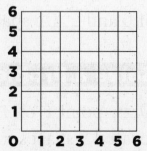

Show What You Know

Below the coordinate grid is a list of points and their ordered pairs. Draw and label the points on the coordinate grid.

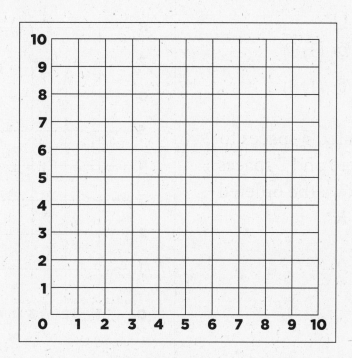

Point A at (1, 5)   Point E at (5, 2)   Point J at (5, 6)

Point B at (2, 4)   Point F at (6, 4)   Point K at (2, 6)

Point C at (1, 3)   Point G at (7, 2)   Point M at (3, 5)

Point D at (2, 2)   Point H at (7, 6)

1. On the coordinate grid above, draw and label point P at the origin.

2. Write the ordered pair for point P.

Point P: (_____, _____)

On Your Own

Connect the points above in this order:

A to B,   B to C,   C to D,
D to E,   E to F,   F to G,
G to H,   H to F,   F to J,
J to K,   K to A.

What picture did you draw?

1. Point Q is located 1 space to the right of zero and 3 spaces up. Which shows the ordered pair for point Q?

   Ⓐ (1, 1) Ⓒ (3, 1)

   Ⓑ (3, 3) Ⓓ (1, 3)

2. Point R is located 3 spaces to the right of zero and 2 spaces up. Which shows the ordered pair for point R?

   Ⓕ (2, 3)

   Ⓖ (3, 2)

   Ⓗ (3, 0)

   Ⓙ (0, 2)

3. Which shows the ordered pair for point K?

   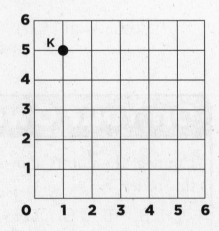

   Ⓐ (5, 1) Ⓒ (1, 5)

   Ⓑ (1, 1) Ⓓ (5, 5)

4. Use the coordinate grid below to draw and label the following points:

   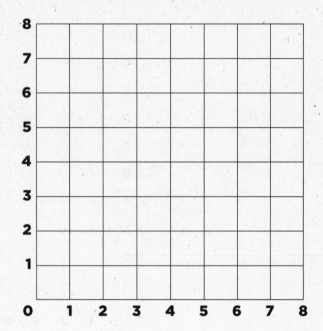

   A (3, 1)   B (7, 1)

   C (7, 4)   D (5, 6)

   　　E (3, 4)

5. Connect each of the points above in order. Then connect point E to point A. What shape did you draw?

   _____

6. **Think Back** On a solid figure the point where three or more edges meet is called a(n) _____ .

Amazing Animals

Animals can do amazing things. Did you know that some frogs can jump 20 times their own length? A flea can jump about 70 times its own height. If you could jump like a flea, you could jump to the top of the Statue of Liberty!

Get Started

One of the students in your class will be the "long jumper." Write his or her name in the chart below. Then follow the steps to see how far that student can jump.

STEP 1 Place a piece of tape on the floor as the start line.

STEP 2 Have the student stand on the start line and jump. Place a piece of tape where he or she lands.

STEP 3 Guess how many feet the student jumped. Write your guess in the chart.

Think:
This page is about 1 foot long from top to bottom.

Student's name:	The student jumped about...

Working with Inches

When you guess how long something is, you **estimate** its length.

To estimate in inches, you can remember that a quarter is about 1 inch wide.

1 inch

A ruler tells you how long different things are. Rulers have marks to show inches and parts of inches. To measure an object with your ruler,

- Start at the "0" end.

- Find the mark on the ruler closest to the end of the object.

1. How long is the lizard? _____ inches

Remember
Sometimes things are not exactly on a mark. You can say "about" when you tell how long they are.

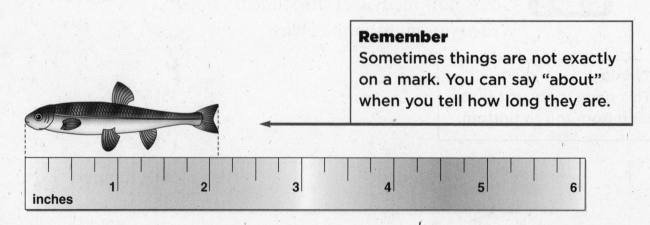

2. About how long is the fish? _____ inches

66   Level C

Practice

Look at each object below. First estimate the length.
Then use your inch ruler to measure.

3. Estimate: _____ inch(es)

Measure: _____ inch(es)

4. Estimate: _____ inch(es)

Measure: _____ inch(es)

5. Estimate: _____ inch(es)

Measure: _____ inch(es)

6. Estimate: _____ inch(es)

Measure: _____ inch(es)

Solve a Problem

7. Find four objects in your classroom that you
think are less than 6 inches long. Estimate the
length of each and then measure.

Object	Estimate	Measure

It's a Fact!

Inchworms are also
called measuring
worms. When an
inchworm is fully
stretched out, it
measures about
1 inch.

Lesson 11: Feet and Inches   **67**

Measuring Feet and Inches

You and a friend entered your frog, Hopper, in a frog-jumping contest. Hopper's longest jump in the contest was between 2 feet and 3 feet. You need to measure Hopper's longest jump, but your ruler is only 1 foot long! Follow the steps below to measure Hopper's longest jump.

STEP 1 Start at the "0" end of your ruler. Make a mark at 12 inches. This is 1 foot.

STEP 2 Line up the "0" end of your ruler with the mark you made. Make another mark at 12 inches.

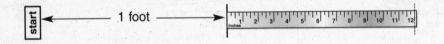

STEP 3 Move the ruler again. Find the inch mark closest to the end of the jump.

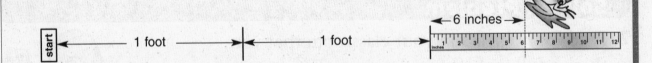

STEP 4 Add to find the total length.

Remember
1 foot = 12 inches

12 inches + 12 inches + 6 inches = _____ inches

1 foot + 1 foot + 6 inches = _____ feet and _____ inches

Show What You Know

1. Find four objects in your classroom that you think are more than 1 foot long. Estimate the length of each. Then use the steps you learned to measure each object. Use the table below to record your data.

Object	Estimate	Measure

2. Compare the four objects you just measured. Write their lengths in order from shortest to longest.

_____ _____ _____ _____

shortest   longest

It's a Fact!
The longest frog jump on record is 33 feet $5\frac{1}{2}$ inches!

On Your Own

Go back to page 65. Use a ruler to measure how far the student jumped. How close was your guess to your measurement?

11 Test Yourself

1. A dog measures 1 foot and 7 inches from head to tail. How many inches long is the dog?

   Ⓐ 8 inches

   Ⓑ 17 inches

   Ⓒ 19 inches

   Ⓓ 37 inches

2. Josh jumped 2 feet. How many inches did he jump?

   Ⓕ 12 inches

   Ⓖ 24 inches

   Ⓗ 36 inches

   Ⓙ 48 inches

3. Jennifer's pen is 7 inches long. How much less than a foot is the length of her pen?

   Ⓐ 3 inches

   Ⓑ 5 inches

   Ⓒ 7 inches

   Ⓓ 9 inches

4. Estimate the length of this picture in inches.

The picture is about _____ inches long.

5. Use your ruler to measure the length of this page from top to bottom.

   This page is about _____ inches long.

6. **Think Back** Use regrouping to find the sum. Show your work below.

   117 + 216 = _____

The Olympics

The Olympics bring athletes from all over the world together to compete in many sports. There are Winter Olympics and Summer Olympics. Did you know that the metric system is used to measure distance in many of the Olympic events? Events like the high jump, the long jump, and the pole vault all use the metric system.

Get Started

Meters and centimeters are units used to measure lengths and distances in the **metric system.**

The distance from the floor to the doorknob on your classroom door is about 1 meter. ⟶

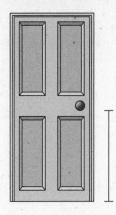

The width of your pinky finger is about 1 centimeter. ⟶

- Look around your classroom and find 3 things that are about 1 centimeter long and 3 things that are about 1 meter long. Complete the table below.

About 1 centimeter long	About 1 meter long

Working with Centimeters

Metersticks and centimeter rulers are used to measure lengths and distances in the metric system. You use a meterstick to measure long objects and a centimeter ruler to measure short objects.

When you measure objects, you can use a centimeter ruler just like you use an inch ruler:

- Start at the "0" end.

- Find the mark on the ruler closest to the end of the object.

1. How long is the pencil? _____ cm

> **Remember**
> You can write 9 centimeters as 9 cm.

2. How long is the nail? _____ cm

It's a Fact!
The motto of the Olympic Games is "Citius, Altius, Fortius," which means "Swifter, Higher, Stronger."

Practice

Look at each picture below. First estimate the
length in centimeters. Then use your centimeter
ruler to measure.

3. Estimate: _____ cm

Measure: _____ cm

4. Estimate: _____ cm

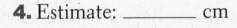

Measure: _____ cm

5. Estimate: _____ cm

Measure: _____ cm

6. Estimate: _____ cm

Measure: _____ cm

Solve a Problem

7. Find three items in your classroom that look like
they are less than 15 cm long. Estimate the length
of each and then measure.

Item	Estimate	Measure

Measuring with a Meterstick

Your class has a day of Olympic Games. Your job is to measure the length of the standing broad jump. The first person to jump is Natalie. She jumps farther than a meter.

To measure Natalie's jump, you can use a meterstick. A meterstick is 100 centimeters long. Follow the steps below to measure Natalie's jump.

STEP 1 Start at the "0" end of your meterstick. Make a mark at 100 centimeters.

|← 100 cm →|

STEP 2 Line up the "0" end of your meterstick with the mark you made. Find the centimeter mark closest to the end of the jump.

|← 100 cm →| 20 cm |

STEP 3 Add to find the total length.

100 centimeters + 20 centimeters = _____ centimeters

1. Find four items in your classroom that look longer than a meter. Estimate the length of each item. Then use the steps you learned to measure with your meterstick. Remember: 1 meter is equal to 100 centimeters.

Item	Estimate	Measure

2. Write the names of the items above in order from shortest to longest.

_____   _____   _____   _____

shortest   longest

3. How much longer is the longest item in your list than the shortest item? Show your work in the space at the right.

_____

4. Use your meterstick to measure the length of your classroom. How long is your classroom in centimeters?

My classroom is _____ cm long.

On Your Own

Go back to page 71. Use a centimeter ruler or a meterstick to measure each item in your table.

1. How long is the crayon?

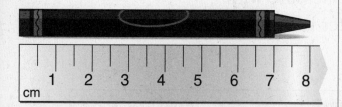

Ⓐ 7 cm   Ⓒ 9 cm

Ⓑ 8 cm   Ⓓ 10 cm

2. How many centimeters are there in 1 meter?

Ⓕ 7   Ⓗ 100

Ⓖ 10   Ⓙ 1,000

3. Barbara has 1 meter of ribbon. If she cuts off 30 centimeters, how much ribbon will she have left?

Ⓐ 20 cm

Ⓑ 30 cm

Ⓒ 70 cm

Ⓓ 130 cm

4. Ask a partner to measure your height in centimeters. Then record the results.

My height = _____ cm

5. Alan measured his height. Alan says that he is 4 meters tall. Do you think Alan is right? Explain why or why not.

_____

_____

_____

_____

6. Think Back Pedro's frog jumped 2 feet and 3 inches in a frog-jumping contest. How many inches did Pedro's frog jump? Show your work.

Pedro's frog jumped _____ inches.

The Newest Coins

The United States Mint makes all of the coins we use every day. In 1999, the Mint began to make new quarters for each of the 50 states. Each state's quarter will have a different design on the back to show something special about that state. Delaware was the first state to have a new quarter, because Delaware was the first state in the United States. Hawaii will be the last state to have a new quarter, because Hawaii was the last state to join the United States.

Get Started

Maria wants to buy a snack for 25¢. She has a new Maryland quarter and some pennies, nickels, and dimes in her pocket. Maria does not want to spend her new Maryland quarter. How could she pay for her snack using pennies, nickels, and dimes?

The chart below shows one way Maria could pay for her snack. Write five more ways.

Dimes	Nickels	Pennies
2	1	0

Working with Money

When you want to buy something, it is important to be able to count how much money you have.

To find the total value of your coins,

- Start with the coins of greatest value.
- Count up to find the total value.

25¢ ⟶ 26¢

Suppose you have these coins:

1. Sort the money so that the coins of greatest value come first. Write what each coin is worth.

25¢							1¢

2. Count up the money, starting with the coins of greatest value. Use the lines below.

25¢ ⟶ 50¢ ⟶ 60¢ ⟶ _____ ⟶ _____ ⟶ _____ ⟶ _____ ⟶ _____

total value

Practice

Write what each coin is worth. Then count up to find
the total value of the coins.

3.

		10¢					

___ → ___ → ___ → ___ → ___ → ___ → ___ → ___

**total
value**

4.

				5¢			

___ → ___ → ___ → ___ → ___ → ___ → ___ → ___

**total
value**

Solve a Problem

5. Bob has 2 quarters, 2 nickels, and 6 pennies.
Max has 1 quarter, 3 dimes, 1 nickel, and
1 penny.

Who has more money? _____

How do you know?

_____

_____

It's a Fact!

The letters "FS" on
the front of this nickel
stand for Felix Schlag.
He designed the
pictures on the front
and back of the nickel.

Making Change

You go to the store to buy a box of birthday candles that costs 39¢. You want to know how much change you should get back if you give the clerk two quarters. Follow the steps below.

STEP 1 Write the amount you give the clerk.

___5___ ___0___ ¢

STEP 2 Write the cost of the birthday candles.

___3___ ___9___ ¢

STEP 3 Subtract the two amounts. Regroup if necessary.

___5___ ___0___ ¢
− ___3___ ___9___ ¢
_____

amount of change ⟶ ___ ___ ¢

STEP 4 Draw the coins you could get back from the clerk.

Does the amount of money in your picture match the amount of change in Step 3?

_____

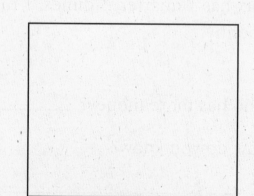

Show What You Know

Use the steps you learned to find out how much change you should get back.

1. You give the clerk 3 quarters.

 59¢

_____ _____ ¢

− _____ _____ ¢

_____ _____

_____ _____ ¢

2. You give the clerk 2 quarters.

 37¢

_____ _____ ¢

− _____ _____ ¢

_____ _____ ¢

3. You give the clerk 1 quarter and 1 dime.

30¢

_____ _____ ¢

− _____ _____ ¢

_____ _____

_____ _____ ¢

4. You give the clerk 3 quarters and 2 dimes.

 89¢

_____ _____ ¢

− _____ _____ ¢

_____ _____ ¢

5. Kelly has 95¢ and wants to buy the raisins and the stamp above. Can she?

_____

_____

On Your Own

On page 77, you showed different ways to make 25¢. How many ways can you show to make $1.00 using coins?

1. What is the value of
2 quarters, 3 dimes,
3 nickels, and 4 pennies?

Ⓐ 99¢

Ⓑ 61¢

Ⓒ 60¢

Ⓓ 44¢

2. Which group of coins has the
least value?

Ⓕ 3 quarters

Ⓖ 5 dimes

Ⓗ 9 nickels

Ⓙ 43 pennies

3. Wendy bought a comb for 79¢.
How much change did she get
if she gave the clerk 3 quarters
and 1 dime?

Ⓐ 16¢

Ⓑ 14¢

Ⓒ 6¢

Ⓓ 4¢

4. Joan has 38¢ in dimes,
nickels, and pennies. Show
3 different groups of coins
Joan might have.

Dimes	Nickels	Pennies

5. Look at the problem above.
If Joan has exactly 8 coins,
what coins does she have?

Write your answer below.

_____

_____

_____

6. Think Back Subtract. Show
your work below.

$$265 - 127 = \underline{\hspace{2cm}}$$

It's About Time

Long ago, people did not have clocks to tell time. In ancient Egypt, people used sundials that cast the sun's shadow to tell the time of day. In ancient China, people burned candles to measure time. The first clocks as we know them were made in Europe 600 years ago. Today, many people use clocks and watches to keep up with their busy lives.

Get Started

Steven checked his watch to see what time he did different things during the day. Draw the hands on the watch faces below to show what Steven's watch looked like at each time.

Start school
8:30 A.M.

Eat lunch
12:00 noon

> **Remember**
> A.M. is before noon.
> P.M. is after noon.

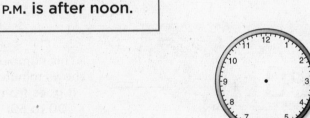

Do homework
3:30 P.M.

Go to bed
9:00 P.M.

- Write the times above in order from earliest to latest.

_____ _____ _____   _____
earliest   latest

Working with Time

We use two types of clocks today. One has a face and hands that point to numbers to tell time.

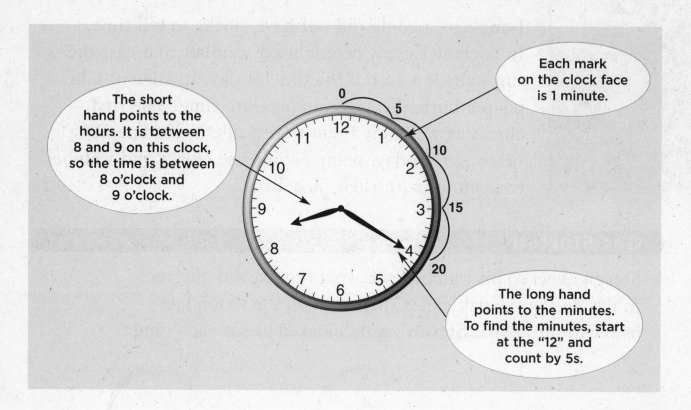

The short hand points to the hours. It is between 8 and 9 on this clock, so the time is between 8 o'clock and 9 o'clock.

Each mark on the clock face is 1 minute.

The long hand points to the minutes. To find the minutes, start at the "12" and count by 5s.

The other clock we use is called a digital clock.

This number shows hours. It goes from 1 to 12.

This number shows minutes. It goes from 00 to 59.

1. Look at both clocks above. Do they show the same time or

different times? _____

2.

Read the time on the clock. Then write the time on the digital clock below.

3.

Read the time on the clock. Then write the time on the digital clock below.

4.

Draw the hands on the clock below to show the time above.

5.

Draw the hands on the clock below to show the time above.

Solve a Problem

6. Think about the time school started today and what time school will be over. Show these times on the clocks below.

School started:

School ends:

Finding Elapsed Time

Suppose you get up at 7:00 A.M. and you are at school by 8:30 A.M. How much time passed between the time you got up and the time you got to school?

To find out, you need to find elapsed time. **Elapsed time** tells you how much time has passed between any two events.

Follow the steps below to find elapsed time.

STEP 1 First count the hours. One hour has passed when the minute hand gets all the way back around to where it started.

7:00 to 8:00 is _____ hour.

STEP 2 Then count the minutes.

8:00 to 8:30 is _____ minutes.

STEP 3 Write the elapsed time:

_____ hour and _____ minutes

Show What You Know

Look at the clocks below. Follow the steps you learned to find the elapsed time.

1. Start   End

P.M.   P.M.

Elapsed time:

____ hours and ____ minutes

2. Start   End

A.M.   A.M.

Elapsed time:

____ hours and ____ minutes

3. Start   End

A.M.   A.M.

Elapsed time:

____ hours and ____ minutes

4. Start   End

P.M.   P.M.

Elapsed time:

____ hours and ____ minutes

5. Start   End

P.M.   P.M.

Elapsed time:

____ hours and ____ minutes

On Your Own

Explain how you can use the steps on page 86 to find the elapsed time in Problems 3 and 4 above.

14 Test Yourself

1. Which clock shows 10:15?

2. How many more minutes are needed to reach 5:00?

Ⓕ 15 minutes

Ⓖ 25 minutes

Ⓗ 30 minutes

Ⓘ 65 minutes

3. Look at the clock below.

What time will it be 10 minutes from the time on the clock?

Ⓐ 2:45   © 12:55

Ⓑ 3:05   Ⓓ 3:45

4. John started his homework at 3:30 P.M. He finished at 4:45 P.M.

Show on the two clocks below the times that John started and finished his homework.

5. How many minutes did John spend doing his homework?

_____ minutes

6. Think Back Count up by 5s to find numbers below.

5, 10, _____, 20, _____, _____

Piggy Banks

Did you ever wonder why we save coins in banks shaped like pigs? Long ago, people in Europe put their spare change in jars made of an orange clay called "pygg." We don't know who made the first bank of pygg clay in the shape of a pig, but we can guess why. Both words sound the same. Today, lots of people put their pennies and other coins in piggy banks.

Get Started

Bart saves pennies, nickels, and dimes in his piggy bank. One day, he opened it to count his money.

First, he took out all of the pennies.
Count up by 1s to find the total value of Bart's pennies.

1¢ _____ 2¢ _____ _____ _____ _____

Next, Bart took out all of his nickels.
Count up by 5s to find the total value of the nickels.

5¢ _____ 10¢ _____ _____ _____ _____ _____

Finally, Bart took out the dimes.
Count up by 10s to find the total value of Bart's dimes.

10¢ _____ 20¢ _____ _____ _____ _____ _____

Working with Number Patterns

A number **pattern** is a list of numbers that repeat or go on in a predictable way. To decide how a number pattern works, you can answer these questions:

A. Do the numbers increase or decrease?

B. By how much each time?

C. What is the **Rule**?

Look at this number pattern: 2, 4, 6, 8, 10

A. The numbers increase from 2 to 10.

B. The numbers increase by 2 each time.

C. The Rule is add 2. You add 2 and 2 to get 4. Then you add 2 and 4 to get 6, and so on.

Now look at this number pattern: 100, 96, 92, 88, 84

A. The numbers decrease from 100 to 84.

B. The numbers decrease by 4 each time.

C. The Rule is subtract 4. You subtract 4 from 100 to get 96. Then you subtract 4 from 96 to get 92, and so on.

Look at this number pattern: 1, 3, 5, 7, 9, 11

1. Do the numbers increase or decrease? _____

2. By how much each time? _____

3. What is the Rule for the pattern? _____

Practice

Study each pattern. Then answer the questions.

4.
| 1, | 4, | 7, | 10, | 13 |

A. Do the numbers increase or

decrease? _____

B. By how much each time?

_____

C. What is the Rule?

_____

5.
| 20, | 16, | 12, | 8, | 4 |

A. Do the numbers increase or

decrease? _____

B. By how much each time?

_____

C. What is the Rule?

_____

6.
| 5, | 10, | 15, | 20, | 25 |

A. Do the numbers increase or

decrease? _____

B. By how much each time?

_____

C. What is the Rule?

_____

7.
| 100, | 90, | 80, | 70, | 60 |

A. Do the numbers increase or

decrease? _____

B. By how much each time?

_____

C. What is the Rule?

_____

Solve a Problem

8. What is the Rule used in the following pattern?

3 dogs, 6 dogs, 9 dogs, 12 dogs

The Rule is _____.

It's a Fact!

Each day, Americans lose or put away about 10 million pennies.

Continuing Number Patterns

Martin and Brenda each have 10¢ in their piggy banks. Brenda puts the same amount in her piggy bank every week. Martin takes out the same amount every week.

You can use what you know about patterns to find out how much money Martin and Brenda will have in their piggy banks each week.

Follow the steps below to continue a number pattern.

		Brenda's Savings	Martin's Savings
STEP 1	Study the pattern.	10¢, 12¢, 14¢, 16¢, . . .	10¢, 8¢, 6¢, 4¢, . . .
STEP 2	Decide if it increases or decreases.	increases	decreases
STEP 3	Decide by how much it increases or decreases each time.	increases by 2¢	decreases by 2¢
STEP 4	Write the Rule for the pattern.	Add 2¢.	Subtract 2¢.
STEP 5	Use the Rule to continue the pattern.	16¢ + 2¢ = _____ 18¢ + 2¢ = _____	4¢ − 2¢ = _____ 2¢ − 2¢ = _____

Show What You Know

Use the steps you learned to continue each pattern.

1. Find the next three numbers in this pattern.

3, 9, 15, 21, _____, _____, _____

Write the Rule. _____

2. Find the next two numbers in this pattern.

30, 27, 24, 21, _____, _____

Write the Rule. _____

3. Find the next three numbers in this pattern.

99, 89, 79, 69, _____, _____, _____

Write the Rule. _____

4. Find the next two numbers in this pattern.

24, 32, 40, 48, _____, _____

Write the Rule. _____

5. What is the first thing to decide when you study a pattern?

_____

_____

On Your Own

Look back at page 89. Suppose Bart finds 3 more pennies, 2 more nickels, and 3 more dimes.

Continue the number patterns to find the total value of each group of coins.

1. Which amount comes next in this pattern?

   35¢, 40¢, 45¢, 50¢, . . .

   Ⓐ 51¢

   Ⓑ 55¢

   Ⓒ 60¢

   Ⓓ 75¢

2. What is the missing number in this pattern?

   80, 70, _____, 50, 40

   Ⓕ 69

   Ⓖ 60

   Ⓗ 51

   Ⓙ 30

3. Which Rule can be used to find the number that comes next?

   4, 8, 12, 16, 20, . . .

   Ⓐ Subtract 4.

   Ⓑ Add 8.

   Ⓒ Add 4.

   Ⓓ Subtract 8.

4. Look at the pattern below. Fill in the next two numbers. Explain how you found your answer.

   12, 15, 18, _____, _____

   _____

   _____

   _____

   _____

   _____

5. Look at the pattern below.

   26, 24, 22, 20, . . .

   What is the next number? _____

   What Rule did you use to find the answer?

   _____

6. **Think Back** Find the sum.

   78 + 14 = _____

Play Ball!

Do you know who invented baseball? Many people believe that Abner Doubleday did in Cooperstown, New York, in 1839. But now history shows that baseball began in New York City in 1845. Alexander Cartwright and his teammates, the Knickerbockers, developed the first rules for the game we know today as baseball. Now it is one of the most popular sports in America. Some people call it "America's favorite pastime."

Get Started

When a baseball team takes the field for a game, there are 9 players.

Circle groups of 2 above.

When you make groups of 2, and there are none left over, you have an **even number.** When you make groups of 2, and there is 1 left over, you have an **odd number.**

- Are there an even number of players on a baseball team or an odd number of players? _____

Working with Odd and Even Number Patterns

To help him remember the sums of even and odd numbers, Brian made a memory table. Look at the addition facts below that Brian used to make his memory table.

1. $4 + 6 =$ _____

Even + Even = _____

Test it! Add two different even numbers: _____ + _____ = _____

2. $2 + 3 =$ _____

Even + Odd = _____

Test it! Add one different even number and one different odd number: _____ + _____ = _____

3. $5 + 4 =$ _____

Odd + Even = _____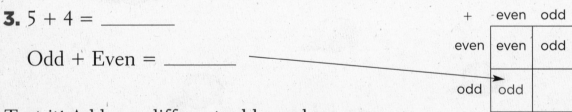

Test it! Add one different odd number and one different even number: _____ + _____ = _____

4. $5 + 3 =$ _____

Odd + Odd = _____

Test it! Add two different odd numbers: _____ + _____ = _____

Look at each addition problem below. Predict whether the answer will be even or odd by circling the word. Solve each problem and check your prediction. Use Brian's memory table to check your work.

5.

16 + 8 = Even or Odd

16 + 8 = _____

6.

25 + 13 = Even or Odd

25 + 13 = _____

7.

32 + 11 = Even or Odd

32 + 11 = _____

8.

19 + 12 = Even or Odd

19 + 12 = _____

9.

58 + 34 = Even or Odd

$$\begin{array}{r} 58 \\ + 34 \\ \hline \end{array}$$

10.

63 + 27 = Even or Odd

$$\begin{array}{r} 63 \\ + 27 \\ \hline \end{array}$$

Solve a Problem

11. Students in Ms. Hill's class and Ms. Engel's class want to play soccer together. There are 13 students in Ms. Hill's class and 15 students in Ms. Engel's class. Can the students make two equal teams so that everyone can play? Explain.

_____

_____

It's a Fact!

The New York Yankees have won more World Series championships than any other baseball team.

Using Odd and Even Number Patterns

During recess, Toby's class formed two equal teams to play tag football. During the game, five students from another class asked if they could play. Is there a way to let all five students play and still keep the number of players on each team equal?

You can use a rule to help you find the answer. Look at the rules below:

- Even + Even = Even

  When you add two even numbers, the sum is always even.

  Example: 4 + 2 = 6

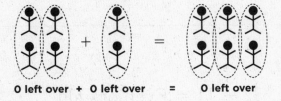

- Even + Odd = Odd
- Odd + Even = Odd

  When you add an even number and an odd number in any order, the sum is always odd.

  Examples: 4 + 3 = 7
  3 + 4 = 7

- Odd + Odd = Even

  When you add two odd numbers, the sum is always even.

  Example: 3 + 3 = 6

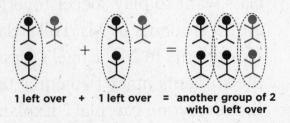

Is there a way to let all five students play? _____

The rules on page 98 also work when you subtract two numbers. Use these rules for Exercises 5–8.

Write whether the sum will be even or odd. Then write an example to show the rule.

1. Even + Even = _____   Example: _____ + _____ = _____

2. Odd + Even = _____   Example: _____ + _____ = _____

3. Odd + Odd = _____   Example: _____ + _____ = _____

4. Even + Odd = _____   Example: _____ + _____ = _____

Write whether the difference will be even or odd. Then subtract.

5. 6 − 2 = Even or Odd

6 − 2 = _____

6. 6 − 3 = Even or Odd

6 − 3 = _____

7. 7 − 4 = Even or Odd

7 − 4 = _____

8. 7 − 3 = Even or Odd

7 − 3 = _____

9. If you are adding or subtracting, and both numbers are even, your answer will always be _____.

10. If you are adding or subtracting two numbers, and only one is even, your answer will always be _____.

On Your Own

Suppose you are adding two odd numbers and an even number.

Will the sum always be even or odd? Draw pictures to help explain your answer.

1. Ellie has an odd number of games. How many games could Ellie have?

Ⓐ 8

Ⓑ 10

Ⓒ 11

Ⓓ 20

2. What will the sum be when any two even numbers are added?

Ⓕ 4

Ⓖ even

Ⓗ 0

Ⓙ odd

3. If you subtract an even number from an odd number, what will your answer be?

Ⓐ odd

Ⓑ even

Ⓒ sometimes even and sometimes odd

Ⓓ 0

4. If you add any two odd numbers, will the sum be even, odd, or sometimes even and sometimes odd?

_____

Explain your answer.

_____

_____

5. Haley had 15 books about sports on her shelf. Did she have an even or an odd number of books?

_____

Draw a picture to explain your answer.

6. Think Back Find the next two numbers in the pattern below.

5, 8, 11, 14, _____, _____

The Inventor of the "Talkies"

Thomas Edison's ideas led to more than 1,000 inventions. His most famous invention is the electric light bulb. However, did you know that he also invented a machine that could record voices and a machine that could record pictures? Talking motion pictures, or "talkies," began in 1913 when Edison put these two machines together.

Get Started

Suppose you could invent a Shape-Changing Machine. Your Shape-Changing Machine would change each shape that was put into it.

Look at the Shape-Changing Machine below. Draw what came out when the last shape was put in.

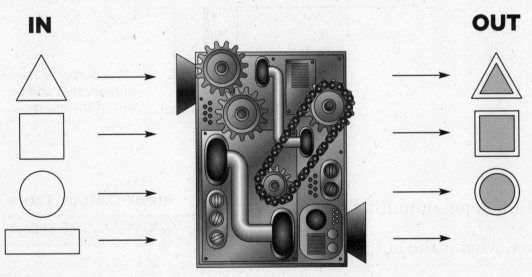

- How does the machine change each shape?

_____

_____

Working with Input-Output Machines

Imagine a machine that uses a Rule to change any number you put into it. If you know the Rule, you always know the number that will come out. The input-output machine below works that way. See the Rule inside. It tells you what will happen.

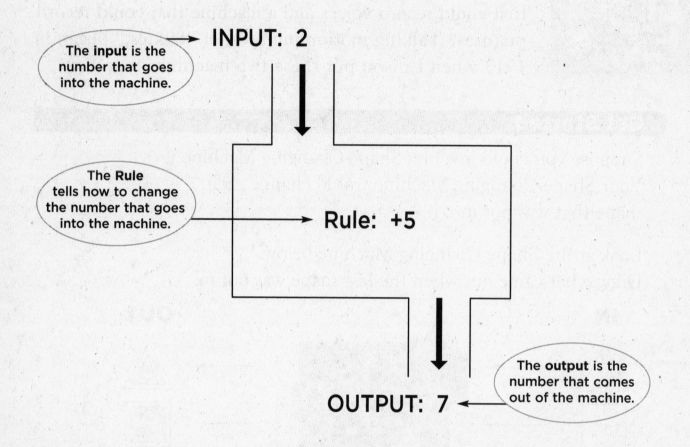

The **input** is the number that goes into the machine.

INPUT: 2

The **Rule** tells how to change the number that goes into the machine.

Rule: +5

The **output** is the number that comes out of the machine.

OUTPUT: 7

1. Look at the input-output table. If the input is 4, what is the output? _____

2. Use the Rule in the input-output machine to complete the table.

Input-Output Table

INPUT	OUTPUT
2	7
3	8
4	9
5	
6	

Practice

Use the Rules in the input-output machines below to complete each table.

3.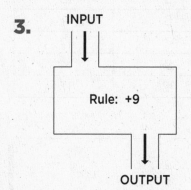

INPUT	OUTPUT
5	14
6	
7	
8	

4.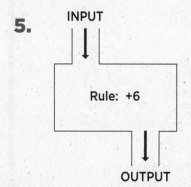

INPUT	OUTPUT
20	15
18	
15	
10	

5.

INPUT	OUTPUT
2	
5	
8	
12	

Solve a Problem

6. Maddie was working with the input-output table for a machine with a Rule of +3. She saw the ouput was 10.

What was the input? _____

It's a Fact!

One of Thomas Edison's laboratories was in Menlo Park, New Jersey. That's why he is sometimes called the Wizard of Menlo Park.

Finding the Rule

Latisha made an input-output machine. She wrote a Rule for the machine that changes each input in the same way.

The table shows the inputs and outputs for Latisha's machine. Follow the steps below to find the Rule.

Input-Output Table

INPUT	OUTPUT
5	10
6	11
7	12
9	14
12	17

STEP 1 Study the table. Look at the first input and the first output.

STEP 2 Guess what the Rule might be.

The Rule might be +5 (5 + 5 = 10), or the Rule might be ×2 (2 × 5 = 10).

STEP 3 Test your guess on the next input.

Does 6 + 5 = 11? _____

Does 2 × 6 = 11? _____

It looks like the Rule is +5, but let's check.

STEP 4 Check the guess that worked on the rest of the inputs.

Does 7 + 5 = 12? _____

Does 9 + 5 = 14? _____

Does 12 + 5 = 17? _____

STEP 5 Write the Rule in the input-output machine.

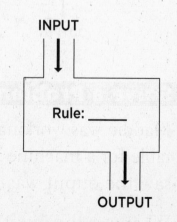

INPUT

Rule: _____

OUTPUT

Use the steps you learned to find the Rule for each of these machines.

1.

INPUT	OUTPUT
1	5
3	7
5	9
7	11

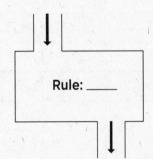

Rule: _____

2.

INPUT	OUTPUT
10	7
8	5
6	3
4	1

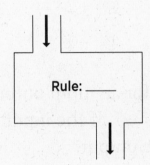

Rule: _____

3.

INPUT	OUTPUT
2	6
4	12
6	18
8	24

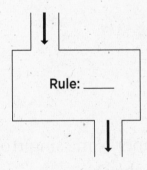

Rule: _____

4. Look at the input-output machine below. Write the missing input.

Input: ____

Rule: × 1

Output: 15

On Your Own

Make up a Rule for an input-output machine. Then fill in an input-output table for your Rule. Give your table to someone and ask them to find your Rule.

1. What is the Rule for the numbers in the table?

INPUT	OUTPUT
4	8
5	9
6	10
7	11

Ⓐ −4   Ⓒ +2

Ⓑ +4   Ⓓ −2

2. The Rule for an input-output machine is −7. If the input is 9, what is the output?

Ⓕ 16

Ⓖ 10

Ⓗ 2

Ⓙ 1

3. What number is missing from the table below?

INPUT	OUTPUT
2	10
4	20
6	30
8	?

Ⓐ 10   Ⓒ 35

Ⓑ 16   Ⓓ 40

4. Look at the input-output machine below. What is the missing input? Explain how you found your answer.

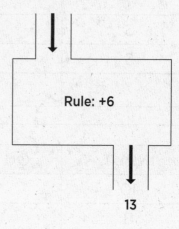

Rule: +6

13

_____

_____

_____

5. Complete the table below. The Rule is +3.

INPUT	OUTPUT
2	5
	9
8	
9	12
15	

6. Think Back Multiply.

$9 \times 9 =$ _____

Super Sleuths

Do you have a favorite storybook sleuth, or detective? Maybe it's Sam the Cat or Tuff Fluff, a rabbit private eye. It's fun to read along as a detective finds important clues. Have you ever solved the mystery before your book's detective did? If you did, then you're a detective too.

Get Started

The numbers in this box can be used to make four **number sentences.** These number sentences make a **fact family:**

6, 5, 11

$$6 + 5 = 11$$

$$5 + 6 = 11$$

$$11 - 5 = 6$$

$$11 - 6 = 5$$

Pretend you are a detective. Use the clues in the fact families below to write the missing numbers.

4, 5, 9

$$4 + 5 = 9$$

$$5 + 4 = 9$$

$$9 - 5 = 4$$

$$\_\_\_ - \_\_\_ = \_\_\_$$

2, 6, 8

$$2 + 6 = 8$$

$$6 + 2 = 8$$

$$\_\_\_ - \_\_\_ = \_\_\_$$

$$\_\_\_ - \_\_\_ = \_\_\_$$

3, 9, 12

$$3 + 9 = 12$$

$$\_\_\_ + \_\_\_ = \_\_\_$$

$$\_\_\_ - \_\_\_ = \_\_\_$$

$$\_\_\_ - \_\_\_ = \_\_\_$$

Working with Open Sentences

One way to think of a number sentence is to think of a balanced scale. The values on both sides of the equals sign must be equal.

An **open sentence** is a number sentence with one of its numbers missing. Like a detective solving a mystery, your job is to find the missing number.

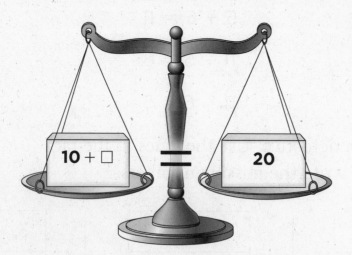

open sentence
a number sentence that has one of its numbers missing

A detective might ask, "What must go in the box so that both sides are equal?" Or a detective might ask, "How much must I add to 10 to get 20?"

1. What number should go in the box above? _____

Study each scale below and answer the questions.

2.

What number belongs in
the box? _____

3.

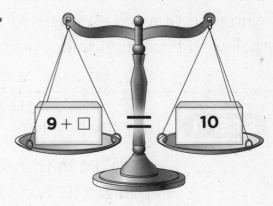

What number belongs in
the box? _____

4.

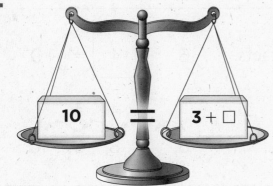

What number belongs in
the box? _____

5.

What number belongs in
the box? _____

Solve a Problem

6. Juan has 15 pounds on his side of a scale.
Bill has 5 pounds on the other side. How
much does Bill have to put on his side to
balance the scale?

_____

It's a Fact!

The Nancy Drew and the Hardy
Boys mystery books have been
popular with youngsters for
more than 70 years.

Solving Open Sentences

Your goal is to read ten mystery stories. You have read six. How many more stories must you read to reach your goal? Follow the steps below to find the number of missing stories.

STEP 1 Study the problem.

Ask, "What must I add to 6 to get 10?"

You have read   Your goal

$$6 + \boxed{} = 10$$

How many more?

STEP 2 Think about your addition facts.

OR

Count up from 6 to 10.

$$6 + \boxed{4} = 10$$

        1   2   3   4

6, 7, 8, 9, 10

STEP 3 Check your answer.

What is the total value of each side?

$$6 + \boxed{4} = 10$$

10 = 10

How many more stories must you read to reach your goal? _____

Show What You Know

Use the steps you learned to find each missing number.

1. $7 + \boxed{} = 12$

_____ = _____

2. $9 + \boxed{} = 15$

_____ = _____

3. $8 + \boxed{} = 11$

_____ = _____

4. $6 + \boxed{} = 13$

_____ = _____

5. What addition fact could you use to find the missing number in the open sentence to the right?

$5 = \boxed{} + 3$

_____

6. What addition fact could you use to find the missing number in the open sentence below?

$6 + \boxed{} = 13$

_____

On Your Own

How could you solve this problem?

$3 + 3 + \boxed{} = 10$

1. Look at the fact family below.

   $5 + 7 = 12$

   $7 + 5 = 12$

   $12 - 5 = 7$

   Which fact is missing?

   Ⓐ $12 + 7 = 19$

   Ⓑ $7 - 5 = 2$

   Ⓒ $12 - 7 = 5$

   Ⓓ $12 + 5 = 17$

2. What number goes into the box to make this number sentence true?

   $8 + \boxed{} = 17$

   Ⓕ 7   Ⓗ 11

   Ⓖ 9   Ⓙ 25

3. What number goes in the box to make this number sentence true?

   $16 = \boxed{} + 7$

   Ⓐ 23   Ⓒ 9

   Ⓑ 11   Ⓓ 8

4. Find the missing number that will balance the scale.

   _____

5. Write the missing number in the box.

   $5 + 3 + \boxed{} = 12$

   Explain how you found your answer.

   _____

   _____

   _____

   _____

6. **Think Back** How many centimeters are there in 1 meter?

   $1\,m = $ _____ cm

A Tasty Mistake

In the 1930s, Ruth Wakefield owned a restaurant named the Toll House Inn. While baking cookies one day, Ruth needed baker's chocolate, but she could not find any. So Ruth cut up a chocolate bar instead and added the pieces to the cookie batter.

The chocolate pieces did not melt like baker's chocolate. They only got softer in the oven. But these cookies, with their "chocolate chips," became her customers' new favorite cookies. That is how chocolate chip cookies were invented—by mistake!

Get Started

Students in Ms. Soto's class wanted to find out how many chocolate chips were in a typical chocolate chip cookie. They opened a new bag and counted the chocolate chips in each cookie. The tally chart below shows the data they collected.

data facts or information

Number of chips	22	23	24	25	26	27	28	29	30	31
Number of cookies		‖‖‖‖ ‖‖‖‖	‖‖‖‖ ‖‖‖‖	‖‖‖‖ ‖	‖‖‖‖ ‖	‖‖‖	‖‖‖‖ ‖			‖‖‖

- What is the greatest number of chocolate chips the class found in a cookie? _____

- How many cookies had more than 25 chocolate chips? _____

Working with Line Plots

Students in Ms. Fox's class sold boxes of cookies to raise money for a community garden near the school. The table to the right shows the data for the class.

Cookie Sales	
Boxes Sold	Number of Students
3	6
4	1
5	3
10	2

The line plot below shows the data from the table. A **line plot** makes it easy to show how data are grouped, or clustered. It uses a number line and X's to show data.

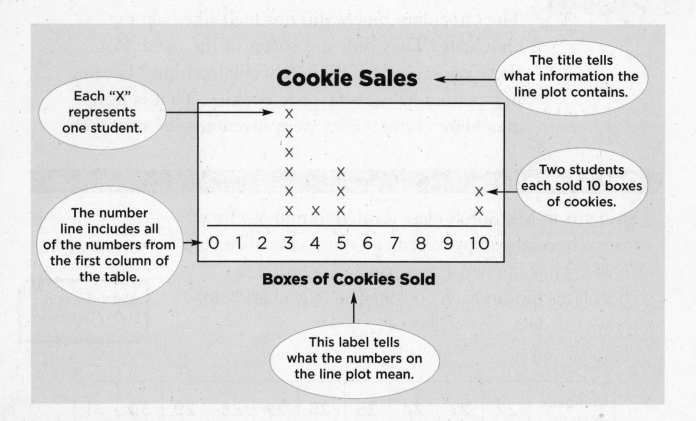

Using the line plot above, answer the following questions.

1. How many students in all sold boxes of cookies? _____

2. How many students sold fewer than 5 boxes of cookies? _____

The line plot below shows the data from the tally chart on page 113. Use the line plot to answer Questions 3–6.

Chocolate Chip Count in Ms. Soto's Class

```
X   X
X   X
X   X
X   X   X   X           X
X   X   X   X           X
X   X   X   X           X
X   X   X   X   X   X           X
X   X   X   X   X   X           X
X   X   X   X   X   X           X
─────────────────────────────────────
23  24  25  26  27  28  29  30  31
```

Number of Chips in a Cookie

3. In the line plot above, what does each X stand for?

_____

4. How many cookies in all did Ms. Soto's class study? _____

5. How many of the cookies had 29 or 30 chocolate chips? _____

Solve a Problem

6. Based on the line plot above, about how many chocolate chips would you expect to find in a chocolate chip cookie? Explain.

_____

_____

_____

It's a Fact!
The chocolate chip cookie is the most popular cookie in America.

Constructing a Line Plot

Ms. Jones offered a large bag of almonds to her class. She told students to take as many almonds as they could hold in one hand. Then she asked them to count the almonds they held. Her findings are shown in the table at the right.

Almonds Students Held in One Hand	
Number of Almonds	Number of Students
8	3
9	6
10	5
12	4
15	1

Use the following steps to show data in a line plot.

STEP 1 Draw a number line that includes all of the numbers in the first column of the table.

Then label the number line. ⟶

```
+---+---+---+---+---+---+---+
8   9  10  11  12  13  14  15
```
Number of Almonds

STEP 2 Draw X's to show the rest of the data.

In this line plot, each X stands for 1 student.

Three students each held 8 almonds.

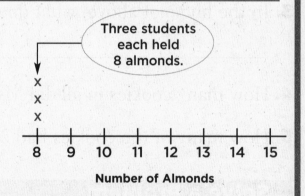

Number of Almonds

STEP 3 Give the line plot a title.

Complete the line plot to show the number of students who held 15 almonds.

Almonds Students Held in One Hand

```
    X
    X   X
    X   X           X
X   X   X           X
X   X   X           X
X   X   X           X
+---+---+---+---+---+---+---+
8   9  10  11  12  13  14  15
```
Number of Almonds

Show What You Know

Abe's grandfather grows tomatoes in his garden. One day, Abe counted the number of tomatoes on each of the tomato plants in his garden. The table at the right shows what Abe counted.

Use the steps you learned to make a line plot of the data in the table.

Tomato Plants	
Number of Tomatoes	Number of Plants
15	4
16	8
19	13
20	7
24	1

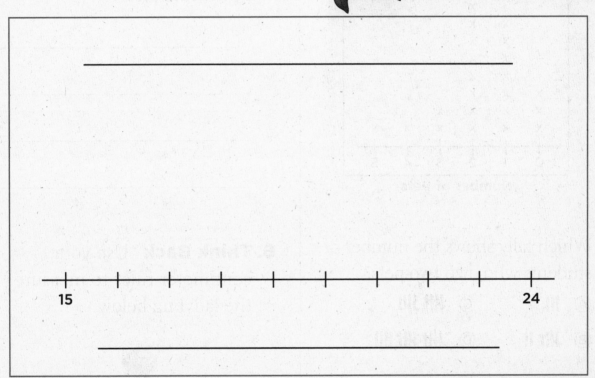

1. How many tomato plants does Abe's grandfather have in his garden? _____

2. Look at your completed line plot. What conclusion can you make about the data?

_____

_____

_____

On Your Own

The next time you eat a chocolate chip cookie, count the number of chocolate chips you see. Then compare the result with the data in the tally chart on page 113.

19 Test Yourself

Abigail surveyed everyone in her class to see how many pets each student has.

Use the line plot to answer Questions 1–5.

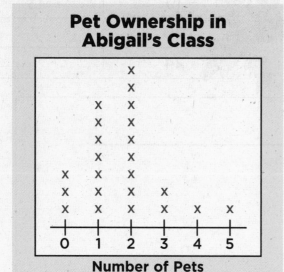

Pet Ownership in Abigail's Class

Number of Pets

1. Which tally shows the number of students who own two pets?

Ⓐ ||| Ⓒ ⫰⫰⫰⫰ ||||

Ⓑ ⫰⫰⫰⫰ || Ⓓ ⫰⫰⫰⫰ ⫰⫰⫰⫰ ||||

2. How many students own 1 pet or 3 pets?

Ⓕ 15   Ⓖ 9   Ⓗ 7   Ⓙ 2

3. How many students in all took part in the survey?

Ⓐ 5   Ⓑ 6   Ⓒ 10   Ⓓ 23

4. How many students in the class own 3 or more pets?

_____ students own 3 or more pets.

5. What conclusions can you make from the survey shown in the line plot?

_____

_____

_____

_____

6. Think Back Use your centimeter ruler to measure the ladybug below.

About how many centimeters long is the ladybug?

_____ cm

118 Level C

© 2005 Options Publishing, Inc.

Up, Up, and Away!

Have you ever imagined taking a ride into space? What about touching a rock from the Moon? Do you wonder what it was like to fly the very first airplane? You can learn all about these things and more at the National Air and Space Museum in Washington, D.C. There you can see real airplanes and space capsules. You can even touch a real moon rock.

Get Started

Suppose 18 classes visited the National Air and Space Museum on Monday, 14 on Tuesday, 20 on Wednesday, 13 on Thursday, and 16 on Friday.

This information, or data, can be hard to understand when it is written in a sentence. One way to make it easier to study is to organize the data in a table. Fill in the data in the table below.

Visitors to the Museum

Day	Number of Classes
Monday	
Tuesday	
Wednesday	
Thursday	
Friday	

- How many classes visited the museum on Thursday? _____

- On which day did 20 classes visit the museum? _____

Working with Bar Graphs

A bar graph lets you compare data by looking at the lengths of different bars. The bars can be vertical, going up and down on the graph. Or the bars can be horizontal, going left and right across the graph.

The bar graph below shows the data from the previous page. The bars are vertical.

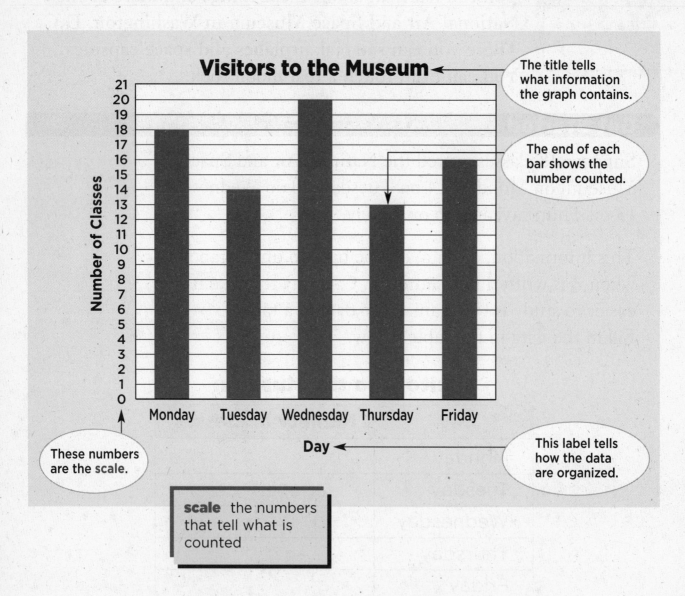

Visitors to the Museum — The title tells what information the graph contains.

The end of each bar shows the number counted.

These numbers are the scale.

scale the numbers that tell what is counted

This label tells how the data are organized.

Using the bar graph, answer these questions:

1. On which day did the most classes visit the museum? _____

2. On which day did 14 classes visit the museum? _____

Practice

The bar graph below shows biographies students read about famous pilots. Study the bar graph. Then answer Questions 3–6.

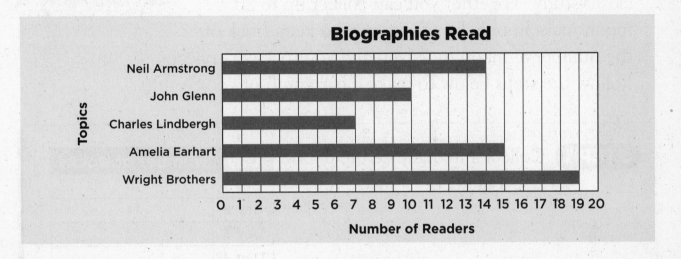

Biographies Read

3. Which two biographies were chosen most?

_____

4. How many more readers read about Amelia Earhart

than about John Glenn? _____

5. Which two biographies were read by 21 children in all?

_____

Solve a Problem

6. Two third grade teams at the Jones School were in a reading contest. Team 1 had students who read about the Wright Brothers or Charles Lindbergh. Team 2 had students who read about Amelia Earhart or John Glenn. Which team had more members?

_____

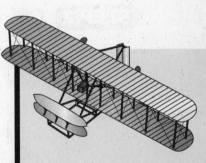

It's a Fact!

On December 17, 1903, the Wright Brothers flew the first powered airplane at Kitty Hawk, North Carolina. The first flight lasted only 12 seconds.

Constructing a Bar Graph

You and a fellow astronaut have been sent to the Moon on a 5-day mission. Your job is to collect moon rocks and bring them back to Earth for closer study. Together you can collect up to 20 moon rocks in one day. Make sure to keep track of the number of moon rocks you and your partner collect. Follow the steps below to make a bar graph of your results.

STEP 1 Collect and organize the data. Use a table.

Day	Number of Rocks
Day 1	10
Day 2	14
Day 3	
Day 4	
Day 5	

STEP 2 Choose a scale for the graph.

This scale counts by twos. →

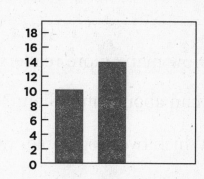

STEP 3 Decide whether the bars will be horizontal or vertical. The bars must be spaced evenly.

STEP 4 Give the graph a title.

Title: _____

STEP 5 Name the labels.

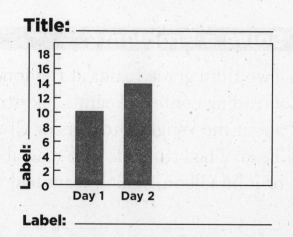

Complete your table on page 122. Then follow the
steps to make a vertical bar graph. Show the number
of moon rocks you and your friend collected.

_____

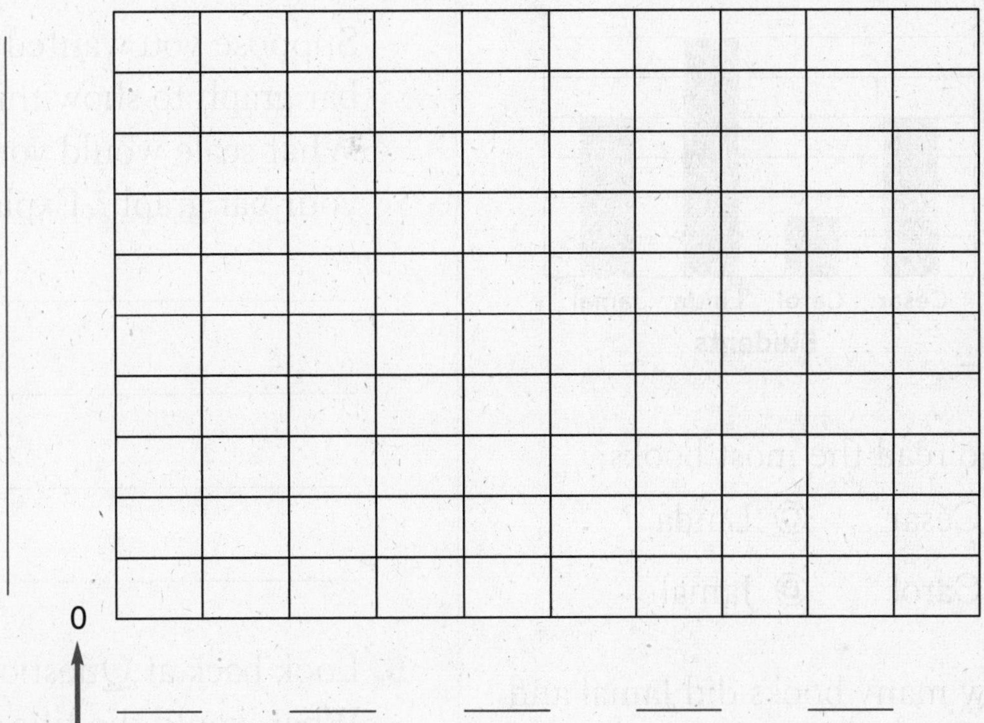

0

Remember
The scale
always starts
at 0.

1. The part of a bar graph that tells you
what the graph is all about is called the

_____ .

2. The part of a bar graph that tells how

many is called the _____ .

On Your Own

Look back at the data on
page 119. Suppose five
more classes visited the
museum each day. Follow
the steps on page 122 to
make a bar graph that
shows the new data.

Four students read books about space. Use the bar graph to answer Questions 1–3.

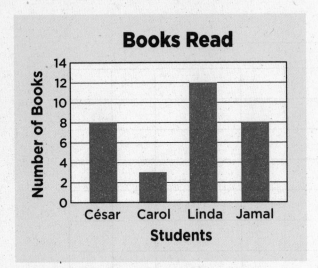

1. Who read the most books?

Ⓐ César   Ⓒ Linda

Ⓑ Carol   Ⓓ Jamal

2. How many books did Jamal and Carol read all together?

Ⓕ 14   Ⓗ 8

Ⓖ 11   Ⓙ 2

3. Which two students read the same number of books?

Ⓐ César and Carol

Ⓑ Carol and Jamal

Ⓒ César and Linda

Ⓓ César and Jamal

4. Over the summer, Maria read 4 books about Mars, 1 book about John Glenn, 1 book about Sally Ride, and 5 books about the moon.

Suppose you wanted to make a bar graph to show this data. What scale would you use in your bar graph? Explain.

_____

_____

_____

_____

5. Look back at Question 4. What would the title of your bar graph be?

_____

_____

6. Think Back What is 764 rounded to the nearest ten?

_____

PROBABILITY 21

Game Time

Do you enjoy playing board games with your friends and your family? Some games, like checkers and chess, require a lot of skill. When you play other games, you cannot win by skill alone. Chance plays a part in these games. There is a chance that you will win. There is also a chance that you will lose!

Get Started

Kim and Hector decide to roll a number cube to see who will start their game. The number cube is marked from 1 to 6. The person who rolls the higher number starts. If they each roll the same number, the person who rolls first gets to start.

Kim rolls first. On the number cubes below, write all of the possible numbers Kim could roll:

Suppose Kim rolls a 4.

- For Hector to start the game, he would have to roll a _____ or a _____.

- Who has a better chance of starting the game—Kim or Hector? _____

Working with Probability

Probability can help you decide how likely it is that an **event** will happen.

Sometimes an event is **certain.** You know that it will happen. On the spinner, it is certain that the pointer will stop on a color.

Sometimes an event is **impossible.** You know that it can never happen. It is impossible that the pointer will stop on Yellow.

It is not certain that the pointer will stop on Blue, but you can say that it is **likely.** You can also say that it is **unlikely** that the pointer will stop on White, but it is not impossible.

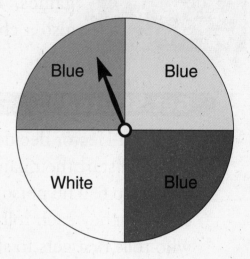

Look at the diagram below. When an event is impossible, it has a probability of 0. When an event is certain, it has a probability of 1.

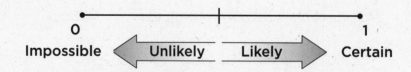

Use the diagram above to complete these sentences:

1. The more likely an event is, the closer the probability is to _____.

2. The more unlikely an event is, the closer the probability is to _____.

Suppose you were asked to pick a ball from each of the four boxes below without looking. Write *unlikely*, *likely*, *certain*, or *impossible* to describe the probability of picking a blue ball from each box.

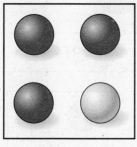
Box A

Box B

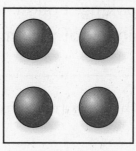
Box C

Box D

3. Box A: _____

4. Box B: _____

5. Box C: _____

6. Box D: _____

Solve a Problem

7. Juan's drawer is filled with six blue shirts and two white shirts. Juan says that if he reaches into the drawer and picks a shirt without looking, he will likely pick a blue shirt. Is Juan correct? Explain your answer.

_____

_____

_____

_____

It's a Fact!

In November of 2003, Mark Allen and Robert Garside received the same number of votes for mayor of a town in Utah. In order to decide who would be mayor, they each had to roll a pair of dice!

Probability as a Fraction

Remember that a fraction is a part of a whole. You can use a fraction to describe the probability that an event will happen.

Suppose you are asked to pick one ball from Box A without looking. Follow the steps below to find the probability of picking a blue ball.

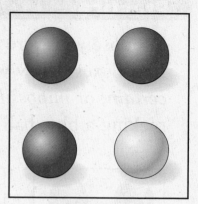

Box A

STEP 1 Find the number of **possible outcomes.**

You are asked to pick one ball. In Box A, there are 4 balls, so there are 4 possible outcomes.

Write "4" below the bar.

← Number of possible outcomes

STEP 2 Find the number of ways the event can happen.

The event is "picking a blue ball." There are 3 blue balls in Box A, so there are 3 different ways to pick a blue ball.

Write "3" above the bar.

← Number of ways event can happen

4 ← Number of possible outcomes

STEP 3 Express your answer.

The probability of picking a blue ball from Box A is 3 out of 4, or _____.

Show What You Know

Follow the steps you learned to find the probability of each event.

1. You spin the pointer once. What is the probability that the pointer will stop on A?

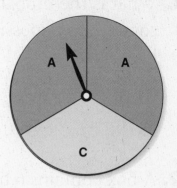

Probability: $\dfrac{\boxed{}}{\boxed{}}$

2. This number cube has six sides labeled 1–6. You roll the number cube once. What is the probability that you will roll a 2?

Probability: $\dfrac{\boxed{}}{\boxed{}}$

3. You spin the pointer once. What is the probability that the pointer will stop on 2?

Probability: $\dfrac{\boxed{}}{\boxed{}}$

On Your Own

Look at Problem 2 above. Find the probability of rolling an even number on the number cube.

21 Test Yourself

1. What is the probability that something impossible will happen?

Ⓐ $\frac{1}{4}$ Ⓒ $\frac{1}{2}$

Ⓑ 1   Ⓓ 0

2. A box contains 6 blue and 4 red crayons. What is the probability of picking a red crayon without looking?

Ⓕ $\frac{4}{10}$ Ⓗ $\frac{6}{10}$

Ⓖ $\frac{4}{6}$ Ⓙ $\frac{10}{10}$

3. What is the probability of the pointer stopping on a shaded part of this spinner?

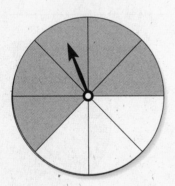

Ⓐ 5 out of 5

Ⓑ 3 out of 8

Ⓒ 5 out of 8

Ⓓ 3 out of 5

4. A set of cards numbered from 1 to 7 is placed facedown. You pick one card without looking. What is the probability of picking an even number?

_____

5. Explain how you found the answer to Question 4.

_____

_____

_____

_____

6. Think Back Shade the model below to show $\frac{1}{6} + \frac{1}{6}$.

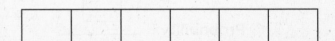

What is $\frac{1}{6} + \frac{1}{6}$? _____

A Dairy Delight

There was a time when people had never heard of ice cream. It wasn't until 1851 that ice cream was first made in America. Today, we can buy hard or soft ice cream and choose from many different flavors. Ice cream is one of America's favorite treats.

Super Scoops Flavors

Vanilla	Mint Chip
Chocolate	Cookie Dough
Strawberry	Chocolate Chip
Rocky Road	Cherry Vanilla

Get Started

On a warm summer day, Stephanie goes to the Super Scoops Ice Cream Stand to buy two scoops of ice cream. She can choose any two flavors. Pick two flavors Stephanie might choose.

Write each flavor on a scoop of ice cream.

Next to the cone, draw a picture to show another way you can arrange the same two scoops of ice cream in the cone.

• How many different ways can you arrange

two scoops in a cone? _____

Working with Arrangements

An **arrangement** is a special way of ordering a collection of things. Stephanie chose two flavors of ice cream. Look at the picture on the right. There are two possible arrangements for two scoops in a cone.

V = vanilla
C = chocolate

Suppose Stephanie wants to buy three different flavors—vanilla, chocolate, and strawberry. How many possible arrangements of three different scoops do you think there will be?

One way to find out is to make an organized list:

- Start with one flavor as the scoop on the top. Write all the ways you could arrange the other scoops under it.

- Repeat this for the rest of the flavors.

	V	V	C	C	—	—
	C	S	—	—	—	—
	S	C	—	—	—	—

1. Complete the list above. How many possible arrangements of 3 different scoops are there? _____

Carlina, David, and Erin are lining up to buy frozen yogurt at Super Scoops. The list below shows one way they can line up.

1st __David__
2nd __Carlina__
3rd __Erin__

2. If David is first in line, how many different ways can Carlina and Erin line up behind him? _____

3. If Carlina is first in line, how many different ways can David and Erin line up behind her? _____

4. Complete the list above. How many different ways in all can the three students line up? _____

Solve a Problem

5. Find all of the 3-digit numbers that can be made using the digits below. Do not use the same digit more than once in a number.

| 6 | 3 | 8 |

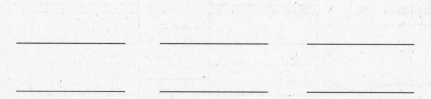

It's a Fact!
The United States makes more ice cream than any other country in the world.

© 2005 Options Publishing, Inc.

Tree Diagrams

Another way to find and show arrangements is to make a **tree diagram.**

Suppose Stephanie went to Super Scoops and bought a cone with 3 scoops of frozen yogurt—vanilla, chocolate, and strawberry. You can make a tree diagram to show all of the possible arrangements of flavors. Follow the steps below.

> **Remember:**
> Every different order is a different arrangement.

STEP 1

Write all of the possible choices for the first flavor.

There are three flavors, so the first scoop can be vanilla, chocolate, or strawberry.

STEP 2

For each of the first flavors, write all of the possible choices for the second flavor.

If the first scoop is vanilla, then the second scoop can be chocolate or strawberry.

STEP 3

Complete the tree diagram.

If vanilla is the first flavor, and chocolate is the second flavor, then strawberry has to be the third flavor.

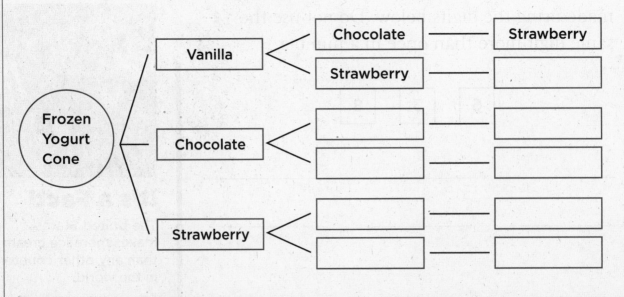

Fill in the tree diagram to show the different ways Carlina, David, and Erin can line up to buy frozen yogurt. Then use the completed tree diagram to answer the questions.

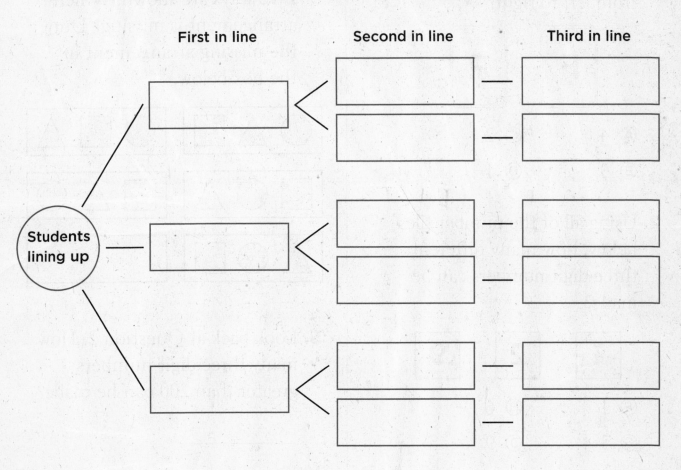

1. List the two ways the three students can line up if Erin is not next to David in line.

Way 1: _____, _____, _____

Way 2: _____, _____, _____

2. Why do you think the diagram at the top of this page is called a "tree diagram"?

_____

_____

On Your Own

Compare the tree diagram you made above with the list on page 133. Find ways they are alike. Find ways they are different.

1. How many different ways can these two cards be arranged from left to right?

ⒶＦ 4   Ⓒ 2

Ⓑ 3   Ⓓ 1

2. Using all of the number tiles below, how many different three-digit numbers can be made?

Ⓕ 1   Ⓗ 6

Ⓖ 3   Ⓙ 9

3. How many ways can three different books be arranged on the shelf below?

Ⓐ 1   Ⓑ 3   Ⓒ 6   Ⓓ 9

4. Michael arranged the shapes below in six different ways. Five ways are shown. Which arrangement is missing? Draw the missing arrangement in the box below.

5. Look back at Question 2. How many three-digit numbers greater than 200 can be made?

_____

6. Think Back Multiply.

$7 \times 8 =$ _____

Glossary

A.M. the time between midnight and noon

arrangement a special way of ordering a collection of things

array things arranged in rows and columns

bar graph a graph that uses bars to show data

certain an event that will definitely happen

congruent figures figures that are the same size and shape

coordinate grid a graph made of horizontal and vertical lines that cross. A coordinate grid is used to locate points.

cube a solid figure with 6 square faces

data facts or information

denominator the bottom number in a fraction. It tells how many equal parts are in the whole.

difference the answer to a subtraction problem

dimension a measure of an object such as length, width, or height

divide to break or separate an object or an amount into equal parts

dividend the number being divided into parts in a division problem

divisor the number that an amount in a division problem is divided by

edge a line segment formed where two faces of a solid figure meet

elapsed time how much time that has passed

estimate to guess

even number a number with 0, 2, 4, 6, or 8 ones. An even number can be divided evenly by 2.

event something that may or may not happen

face one of the flat sides of a solid figure

fact family a group of facts that use the same numbers when they are added and subtracted. Here is an example of a fact family:
$3 + 4 = 7, 4 + 3 = 7, 7 - 4 = 3, 7 - 3 = 4.$

factor the numbers that are multiplied to get a product

fraction a part of a whole

horizontal going left and right

impossible an event that cannot happen

likely an event that will probably happen

line of symmetry a line that divides a figure into two halves that look exactly alike

line plot a diagram that uses X's above a number line to show data

metric system a measuring system that includes units such as meters and centimeters

multiply to add a number to itself as many times as is shown by another number

net a flat pattern that you can fold to make a solid figure

number line a line that shows numbers arranged from least to greatest

number sentence a sentence that uses numbers, an operation sign, and an equals sign. An example of a number sentence is $2 + 3 = 5.$

numerator the top number in a fraction. It tells how many parts of the whole you are talking about.

odd number a number with 1, 3, 5, 7, or 9 ones. An odd number cannot be divided evenly by 2.

open sentence a number sentence that is missing a number. $2 + ? = 5$ is an example of an open sentence.

ordered pair a pair of numbers that names where a point is located on a map or grid

origin the point where the two number lines of a coordinate grid meet

pattern something that repeats or goes on in a way that you can predict

place value the value of a digit based on its place in a number. For example, the place value of the digit 5 in 350 is 50.

plane figure a figure with two dimensions

P.M. the time between noon and midnight

possible outcome a result that could happen in a probability experiment

probability the chance of an event happening

product the answer that you get when you multiply numbers

quotient the answer that comes from solving a division problem

rectangle a 4-sided figure with opposite sides that are the same length

rectangular prism a solid figure with six sides. Each side is a rectangle.

regrouping to take an amount from one place value and rename it as a different place value. Examples: 1 group of ten = 10 ones, and 10 ones = 1 group of ten.

round to find a value that is close to the real value but easier to work with

rule a direction that you follow to get from one number in a pattern to the next number in the pattern

scale the numbers that tell what is counted on a graph

solid figure a figure with three dimensions

square a figure with 4 equal sides and 4 equal angles

square pyramid a solid figure made up of one square face and four triangular faces

sum the answer that you get when you add numbers

tree diagram a picture that shows the possible outcomes of an event

triangle a plane figure with 3 sides

unlikely an event that probably will not happen

vertex the point where three or more edges of a solid figure meet

vertical going up and down

Number Line

A number line can help you add and subtract. To add 2 and 4, put your finger on the 2. Then move it to the right 4 spaces. Your finger should end up on 6, so you know that 2 + 4 = 6.

To subtract 3 from 5, put your finger on the 5. Then move it to the left 3 spaces. Your finger should end up on 2, so you know that 5 − 3 = 2.

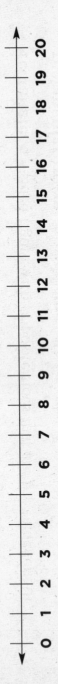

0 1 2 3 4 5 6 7 8 9 10 11 12 13 14 15 16 17 18 19 20

Place Value

Place value tells you the amount that each digit in a number stands for.

5
ones

5 ones = 5

215 = 2 + 1 + 5
hundreds tens ones

2 hundreds = 200

1 ten = 10

5 ones = 5

15 = 1 + 5
tens ones

1 ten = 10

5 ones = 5

3,215 = 3 + 2 + 1 + 5
thousands hundreds tens ones

3 thousands = 3,000

2 hundreds = 200

1 ten = 10

5 ones = 5

Multiplication Table

A multiplication table can help you practice the basic multiplication facts. To use a multiplication table, follow these steps:

- Find one of the numbers being multiplied in the first column.
- Find the other number being multiplied in the top row.
- Find the place where the row and column meet; that number is the product.

Example: $2 \times 5 = 10$

x	0	1	2	3	4	5	6	7	8	9
0	0	0	0	0	0	0	0	0	0	0
1	0	1	2	3	4	5	6	7	8	9
2	0	2	4	6	8	10	12	14	16	18
3	0	3	6	9	12	15	18	21	24	27
4	0	4	8	12	16	20	24	28	32	36
5	0	5	10	15	20	25	30	35	40	45
6	0	6	12	18	24	30	36	42	48	54
7	0	7	14	21	28	35	42	49	56	63
8	0	8	16	24	32	40	48	56	64	72
9	0	9	18	27	36	45	54	63	72	81

Rulers

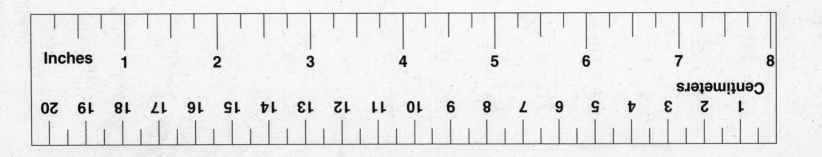